CONTENTS

1. INTRODUCTION:

Welcome to a compelling journey through the evolutionary history of algae. As we embark on this exploration, we confront a fundamental question that has intrigued biologists for decades: How did the diverse and complex world of algae evolve, and what is its connection to cyanobacteria[1]? Unlike many evolutionary paths, the journey of algae unfolds not as a linear progression but as a rich tapestry, intricately woven with multiple origins converging to form the diverse group known as algae.

Central to our narrative is the fascinating concept of endosymbiosis[2]. More than a theoretical construct, endosymbiosis emerges as the pivotal mechanism driving the evolution of algae and, indeed, all complex life. It tells a story of symbiotic relationships, where disparate lifeforms unite, leading to remarkable evolutionary leaps. Within these partnerships, once independent organisms evolve into an integral part of their hosts, reshaping their own destinies and the course of life on Earth.

This book explores the multifaceted nature of endosymbiosis, illustrating how it transcends simple survival to foster intricate networks of cooperation. These networks underpin the transformation from the

---

[1] **Cyanobacteria**: Photosynthetic bacteria, pivotal in oxygen production and nitrogen fixation, present in various habitats, existing for over 2.5 billion years, can form toxic blooms.

[2] **Endosymbiosis**: When one organism lives inside another, and both benefit. This led to complex cells (eukaryotes) evolving from simpler ones (prokaryotes).

simplicity of prokaryotic[3] cyanobacteria to the complexity of eukaryotic[4] algae. Our exploration is not confined to theoretical musings; it is grounded in the empirical evidence uncovered by modern microbiology, offering a comprehensive view of the algal world.

Through a blend of historical insights and current scientific understanding, we will navigate the myriad strategies and mechanisms that have fueled this evolutionary marvel. Our journey will showcase how the fusion of simplicity and complexity through endosymbiosis has crafted algal life's vast and varied tapestry.

Join us as we delve into this profound exploration, where each chapter unfolds the intricate story of algal evolution. Here, microscopic partnerships not only etch the history of algae but also reveal the intricate interplay of life's forces, uniting the primal with the complex in a dance of evolutionary innovation.

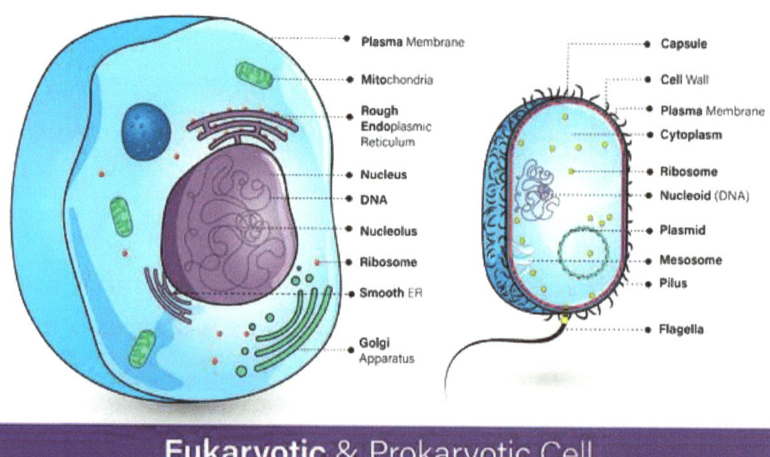

Figure 1.1 Credit maxineu.bio

[3] **Prokaryotic**: Cells that do not have a nucleus or other organelles. Bacteria are an example of prokaryotic organisms, their DNA floats freely inside the cell.
[4] **Eukaryotic**: Cells that have a nucleus and other membrane-bound organelles. Plants, animals, fungi, and protists are made of eukaryotic cells.

2. THE HISTORY OF ENDOSYMBIOTIC THEORY AND ITS IMPORTANCE IN ALGAL EVOLUTION

2.1 Background

Before we embark on the journey of algal evolution, I would like to take you through the history of the development of endosymbiotic theory[5]. This theory is a cornerstone to understanding algal evolution. At its core, endosymbiotic theory encompasses the profound relationships between organisms, unlocking the secrets of how life on Earth has evolved and diversified over countless eons. If we want to understand the significance of this theory, it is essential to define the key concepts.

Endosymbiosis, a term derived from the Greek words "endo" ("inside") and "symbios" ("to live together"), describes a captivating phenomenon in which one organism takes up residence within another, forming a mutually beneficial or symbiotic relationship. In this intricate dance of coexistence, one organism becomes the host while the other resides within it, often within specialized structures or organelles. The fundamental premise of endosymbiosis is that both the host and the internalized organism derive benefits from this intimate association. However, it is essential to note that not all such associations have a

---

[5] **Theory:** A well-substantiated explanation of some aspect of the natural world, based on a body of evidence that has been repeatedly confirmed through observation and experimentation. It provides a framework for understanding observations and making predictions about natural phenomena.

harmonious outcome, and they can range from parasitic to mutually beneficial, ultimately shaping the course of evolution. (Wernegreen, 2012)

In less than a decade, endosymbiotic theory went from a mocked hypothesis to holding a prominent place in evolutionary biology. It challenges conventional perspectives on the gradual accumulation of mutations and natural selection as the sole drivers of evolution. While these processes undoubtedly play pivotal roles in the evolution of life forms, including algae, endosymbiosis introduces an additional layer of complexity, emphasizing the role of chance interactions and unexpected mergers between distinct organisms. The biological significance of endosymbiosis is evident in the complexity of eukaryotic cells, which harbor membrane-bound organelles[6], separate, specialized structures responsible for various cellular functions.

2.2 From Crap to Theory

Figure 1.1 Lynn Margulis at the III Congress for Science Communication in La Coruña, Spain, captured during her presentation on November 9, 2005. (Trustees of Dartmouth College)

The journey to our current comprehension of endosymbiosis is multifaceted and sometimes contentious, marked by pivotal contributions from renowned scientists who challenged conventional wisdom. One of these pioneers was evolutionary biologist Lynn Margulis, whose groundbreaking hypothesis on the origins of organelles within eukaryotic cells

---

[6] **Membrane-bound organelles**: Structures within a cell that are enclosed by a membrane, each with a different function. Examples include the nucleus and mitochondria. Prokaryotic cells, like bacteria, do not have these.

stirred controversy and skepticism in the scientific community. The response she received to her grant application encapsulated the sentiment of the time, which bluntly stated, "Your research is crap. Don't bother applying again." Imagine the discouragement of receiving such a dismissive assessment for a grant application that sought to fund one's life's work. (Gray, 2017) (Sagan D. , 2012)

Margulis dared to suggest that genetics alone did not provide the complete narrative of evolution. In her influential paper, "On the Origins of Mitosing Cells," published in 1967, she advanced the notion that while genetic variation and natural selection certainly played essential roles in evolution, chance events and interactions between unrelated cells might have provided the catalytic spark for the transition from simple prokaryotic cells to the intricate molecular machinery of eukaryotes. (Sagan L. , 1967) Her concept, however, faced significant resistance when she initially presented it. The paper was rejected 15 times before being accepted in the Journal of Theoretical Biology. (Knoll, 2012)

In this groundbreaking paper, Margulis proposed that organelles like chloroplasts[7] and mitochondria might not have originated through the gradual process of iterative mutation and natural selection. Instead, she offered that these organelles resulted from extreme symbiotic collaboration. Her theory suggested that ingested prokaryotic cells failed to be digested by their host and evolved into organelles through mutually beneficial relationships. Margulis postulated that organelles such as chloroplasts, mitochondria, and even external appendages like flagella[8] might have their roots in fortuitous symbiotic encounters. (Sagan L. ,

---

[7] **Chloroplasts**: Organelles found in plant and algal cells responsible for photosynthesis, converting light energy into sugars, containing their own DNA, similar to bacteria.

[8] **Flagella**: Long, whip-like structures that protrude from the cell body and are used for movement or sensing the environment in various microorganisms.

1967) Subsequent research cast doubt on the hypothesis that flagella originated via endosymbiosis; however, researchers found compelling evidence supporting organelles' endosymbiotic origins. (Jing Yan Li, 2005) (Wernegreen, 2012)

Margulis was not the first to propose such an idea. In 1905, Russian botanist Konstantin Mereschkowski put forward a theory in "The Nature and Origins of Chromatophores in the Plant Kingdom." He drew from the observations of botanist Andreas Schimper, who noted the striking resemblance between the separated organelles of chloroplasts in plants and free-living cyanobacteria in appearance. Mereschkowski postulated that "the theory of symbiogenesis is a theory of selection relying on the phenomenon of symbiosis." (William Martin K. K., 2010)

However, his work, along with the contributions of others, failed to capture the widespread scientific interest of that era. It wasn't until the advent of electron microscopy and the discovery that specific organelles possessed distinct genetic material that the scientific community seriously considered the possibility of symbiotic origins for these organelles. This new era of innovation in microbiology and microscopy played a pivotal role in shaping Margulis's development of the endosymbiotic theory.

Subsequently, in the 1970s and 1980s, scientists could put the endosymbiotic theory to the biochemical test as genetics advanced and genetic sequencing emerged. Microbiologists Carl Woese and George Fox pioneered using small subunit ribosomal RNA (SSU rRNA) gene sequences to decipher evolutionary relationships among species. This pioneering work led to a profound shift in our understanding of phylogeny and the revelation of the archaea domain of life[9]. (Garrett, 2014)

Figure 4. George E. Fox
(University of Houston, n.d.)

SSU rRNA is a universal marker found in all living cells, making it an invaluable tool for exploring evolutionary relationships. Woese and Fox harnessed this marker to uncover the shared evolutionary ancestry between mitochondria and prokaryotic bacteria. SSU rRNA acts as a molecular clock, enabling scientists to estimate the timing of critical events in the phylogenetic[10] evolution of a species. (Garrett, 2014)

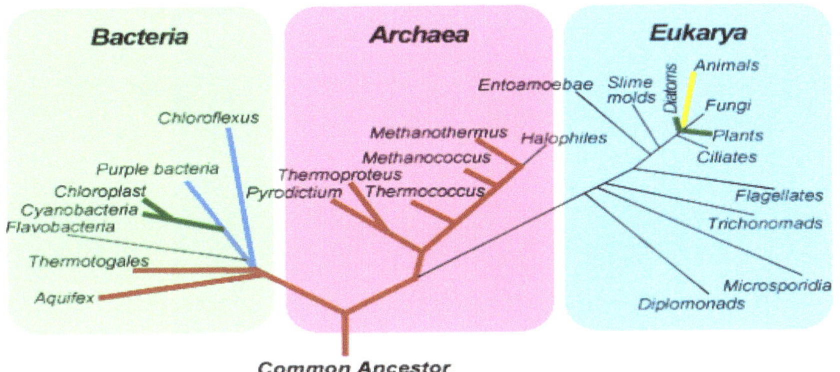

Figure 2-5.. Graphic of the 3 domains of life and how they are related (Bickford, 2013)

---

[9] **Domain of life:** The highest taxonomic rank in the biological classification system, grouping all forms of life into three broad categories: Bacteria, Archaea, and Eukarya.

[10] **Phylogenetic**: The study of the evolutionary history and relationships among different groups of organisms, like figuring out a family tree for all life forms.

Figure 2-6. Linda Bonen (University of Ottawa , n.d.)

This research established a foundation for further exploration into the origins of red algae. Linda Bonen and Ford Doolittle of the University of Nova Scotia utilized ribonuclease digestion[11] of 16S and 18S rRNAs[12] to analyze oligonucleotide[13] sequences, comparing these sequences across various prokaryotes and the red alga *Porphyridium*, laying the groundwork for understanding the evolutionary origins of red algal chloroplasts[14]. Their findings suggest that these chloroplasts likely evolved from endosymbiotic prokaryotes, as indicated by the significant sequence similarities between red algal chloroplast 16S rRNA and bacterial 16S rRNA, specifically *Bacillus subtilis* and, to a lesser extent, blue-green algae[15]. This work supports the hypothesis that the chloroplasts found in major eukaryotic algal groups may have originated from multiple

Figure 2.7. Ford Doolittle (Dalhousie University)

---

[11] **Ribonuclease digestion:** A process where ribonuclease enzymes cut RNA molecules into smaller fragments, often used to study RNA structure and sequence in molecular biology.

[12] **16S and 18S rRNAs:** Components of the small subunit of prokaryotic and eukaryotic ribosomes, respectively, essential for protein synthesis and widely used in phylogenetics.

[13] **Oligonucleotide**: A short strand of nucleic acid, typically DNA or RNA, made up of a few nucleotides, used in genetic testing and research.

[14] **Chloroplasts**: Organelles in plant and algae cells where photosynthesis occurs, converting sunlight to energy, containing their own DNA, descended from ancient photosynthetic bacteria.

[15] **Blue-green algae**: Scientifically know as cyanobacteria, are photosynthetic bacteria often mistaken for true algae due to similar functions but are prokaryotic, lacking distinct nuclei and organelles.

independent lines of prokaryotic, photosynthetic endosymbionts[16] (Linda Bonen, 1975).

2.3 Multilayer Endosymbiosis

Figure 2-8. Patrick J. Keeling By Nwiebe - Own work

Researcher Patrick J. Keeling made significant strides in advancing our understanding of multilayer endosymbiosis, specifically through his extensive work on *apicomplexan* parasites (Varsha Mathur W. K., 2021) (Waldan K Kwong, 2019) (Patrick J. Keeling V. M., 2021). These intriguing microorganisms have provided many insights into complex, multi-layered symbiotic relationships (Varsha Mathur M. K., 2019). Keeling made a remarkable discovery that centered on the organelle known as the apicoplast[17]. This organelle, found in apicomplexan parasites, is a secondary plastid[18] with origins tracing back to red algae, a division of photosynthetic eukaryotes known for their vibrant hues (Nobuko Arisue, 2015).

To truly appreciate the significance of Keeling's findings, it's essential to understand the profound implications of the apicoplast's

---

[16] **Endosymbiont**: an organism that lives inside the body or cells of another organism in a mutually beneficial relationship, often essential to their host's survival. In the context of this book, the word will be used for any entity which can trace its origin to endosymbiosis.

[17] **Apicoplast**: A unique, non-photosynthetic chloroplast found in the cells of most apicomplexan parasites, like malaria, involved in essential metabolic processes.

[18] **Secondary plastids**: Organelles derived from the engulfment of algae containing primary plastids, found in some eukaryotic cells, crucial for photosynthesis in diverse lineages.

ancestry. Plastids[19] are important organelles responsible for photosynthesis in many eukaryotic cells, and their origins have long been a subject of scientific curiosity (Varsha Mathur W. K., 2021) (Waldan K Kwong, 2019). In the case of the apicoplast, Keeling's research unearthed a connection that reaches back millions of years, unraveling the evolutionary tapestry of these microscopic parasites (Nobuko Arisue,

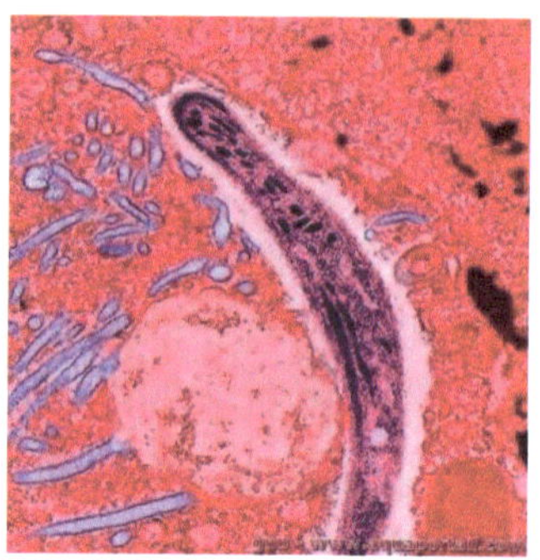

Figure 2-9. *Plasmodium bergei*, an apicomplexan, the agent that causes malaria (AquaPortail, 2023)

2015). In simple terms, millions of years ago, the ancestors of the apicomplexan parasites may have eaten red algae. The red algae, rather than being digested, survived within the parasite. Over millions of years, the plastid made a staggering leap from photosynthetic energy production to producing a wide range of compounds necessary for the parasite to be able to invade a host and set up its parasitic relationship (Varsha Mathur W. K., 2021) (Varsha Mathur M. K., 2019) (Nobuko Arisue, 2015).

Utilizing advanced techniques such as phylogenetic mapping[20], Keeling embarked on the ambitious task of creating phylogenetic trees. Like the branches of life's evolutionary family tree, these trees aimed to showcase the intricate connections between host eukaryotic cells and their organelles (Patrick J. Keeling V. M., 2021) (Varsha Mathur M. K., 2019).

---

[19] **Plastids:** Organelles found in the cells of plants and algae, responsible for photosynthesis, storage, and synthesis of chemical compounds within the cell.
[20] **Phylogenetic mapping:** A method used to reconstruct evolutionary relationships by comparing genetic, morphological, or molecular features of different organisms.

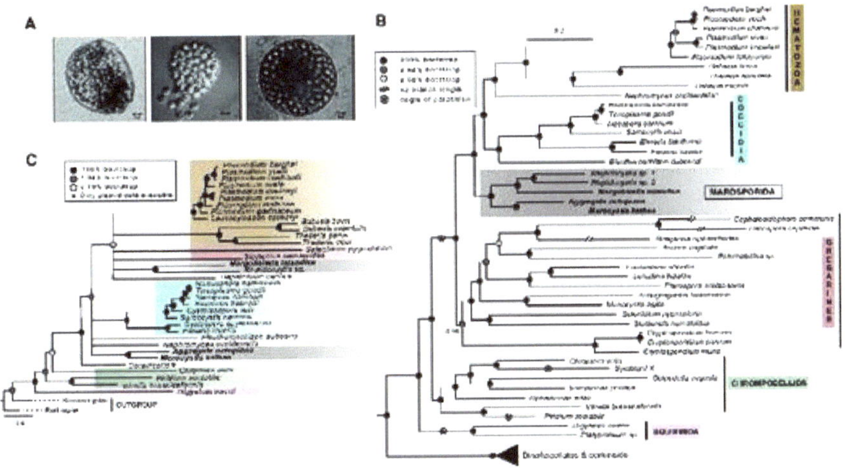

Figure 2-10. Phylogenetic tree created by Keeling (Varsha Mathur W. K., 2021)

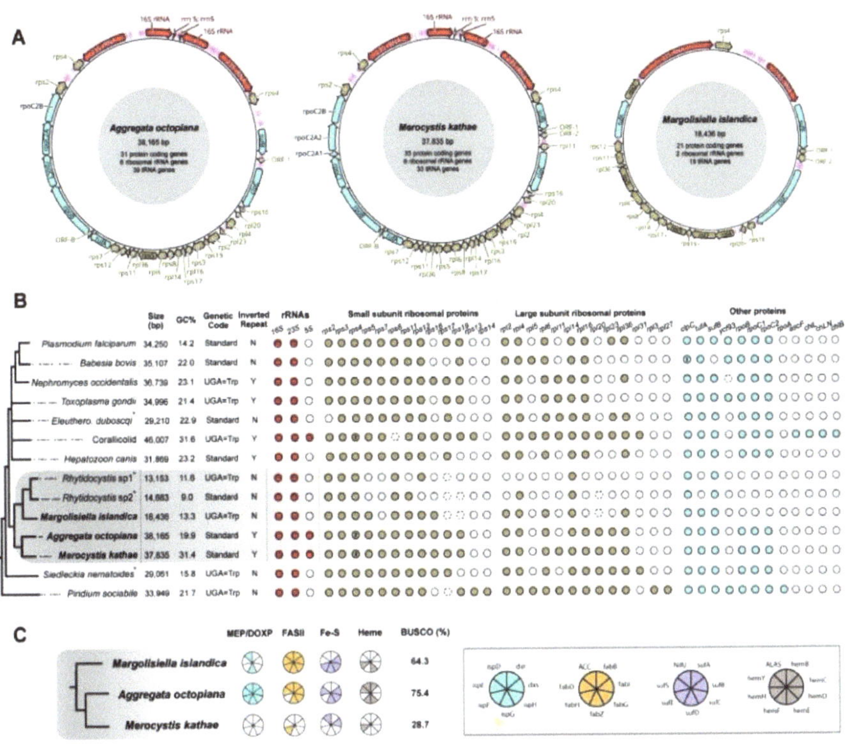

In particular, Keeling's work illuminated the convoluted and fascinating relationship between apicomplexan parasites and the organelles that reside within them (Varsha Mathur W. K., 2021) (Nobuko Arisue, 2015).

The evolutionary history of these parasites unfolded through the lens of phylogenetic mapping. Keeling's research allowed us to understand the ancient associations between host cells and their organelles forged through the intricate dance of endosymbiosis. The phylogenetic trees vividly depicted how endosymbiosis has shaped the complex relationships among these microscopic organisms (Patrick J. Keeling V. M., 2021) (Varsha Mathur M. K., 2019) (Nobuko Arisue, 2015).

Figure 2-12. Dr Jan Janouskovec PhD University of Southampton

Researcher Jan Janouskovec worked with Keeling to delve deeper into secondary and tertiary endosymbiosis by studying alveolites, a diverse clade of single-celled organisms belonging to the *Alveolata superphylum* of coral. Alveolates have captured scientists' fascination for their vast array of characteristics and ecological roles, making them an intriguing subject of study in endosymbiosis. This group of microorganisms comprises various lineages, including dinoflagellates, apicomplexans, and ciliates, each with unique features and functions. For instance, dinoflagellates play a pivotal role in marine ecosystems, serving as primary producers that influence the dynamics of aquatic food chains. Some dinoflagellates are also notorious for forming harmful algal blooms, the consequences of which can be detrimental to marine ecosystems (Jan Janouskovec, 2010).

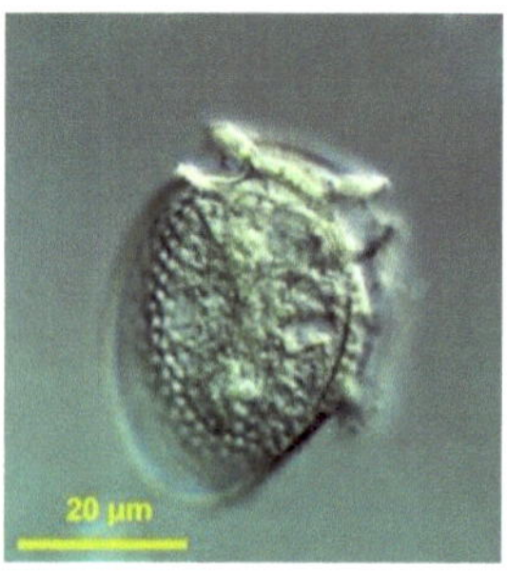

Figure 2-13. *Dinophysis acuminata* dinoflagellate Credit: fjouenne http://planktonnet.awi.de/

Apicomplexans, parasites responsible for malaria, highlight the importance of understanding these diverse organisms' biology and evolutionary history not just for ecological research but also because of the implications for medical and public health.

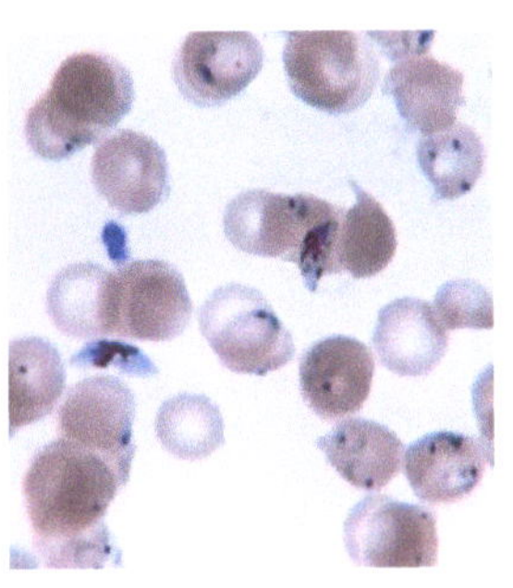

Figure2-14. *Plasmodium falciparum* attacking blood cells Credit: TimVickers

Genetic analysis of these microorganisms unveils a rich tapestry of endosymbiosis, reflecting a complicated phylogeny within protists [21] . This intricate genetic web hints at the dynamic relationships of organisms with their organelles and the myriad sources from which they gain these essential components (Jan Janouskovec, 2010).

Janouskovec's research reveals the complex secondary and tertiary endosymbiosis processes through meticulous examination of their genetic makeup. These microorganisms serve as living archives of ancient endosymbiotic events, providing insights into how organelles have evolved from diverse sources over geologic time. This genetic treasure trove allows us to piece together the puzzle of the evolutionary forces that have shaped the various characteristics and functions observed in algae today (Jan Janouskovec, 2010).

<u>2.4 Sharing Genetics</u>

Researcher Jan Andersson further extended our knowledge of various aspects of endosymbiosis, from understanding the exchanges between host cells and their symbionts[22] to deciphering how the host cells

---

[21] **Protists**: A diverse group of eukaryotic microorganisms, which are neither plants, animals, nor fungi, and often live in aquatic environments.
[22] **Symbiont**: An organism that lives in close association with another different organism, engaging in a symbiotic relationship that can be beneficial, harmful, or

regulate their genes. This intricate web of interactions is a testament to the ability of endosymbiosis to shape the biological world. One of the key revelations from Andersson's research is the realization that horizontal gene transfer[23] is not just a theoretical concept but a tangible phenomenon occurring in these relationships (Vincent Daubin, 2016). In horizontal gene transfer, genes from one organism are transferred to another, often

Figure 2-15. Jan Andersson Uppsala Universitet

unrelated organism (Filip Husnik, 2017). This exchange of genetic information can have profound implications for the evolution of both the host and symbiont (Finan, 2002) (Debashish Bhattacharya T. N., 2008) (Andersson, 2009).

Andersson's groundbreaking work has showcased that, within endosymbiosis, genetic exchange is not limited to vertical inheritance[24], where genes from a parent pass to their offspring. Instead, these organisms engage in a horizontal genetic exchange, sharing genetic material that challenges traditional notions of inheritance and evolution. The genetic information of symbionts becomes integrated into the host genome, creating a dynamic genetic landscape where the boundaries between different organisms blur (Andersson, 2009).

---

neutral to the host. In the context of this book, symbiont is used mainly to refer to organisms and entities within other organisms.

[23] **Horizontal gene transfer:** The movement of genetic material between organisms in a manner other than traditional reproduction, crucial for genetic diversity and evolution.

[24] **Vertical inheritance:** The transmission of genetic material from parent to offspring during the process of reproduction, primarily through sexual or asexual means.

This phenomenon is particularly significant in the context of secondary endosymbiosis. During these events, a eukaryotic host engulfs another eukaryotic cell, integrating the engulfed cell and its organelles into the host's biology. This process often results in an intricate fusion of genetic information, where the host and engulfed cell become intertwined in a genomic dance that shapes their shared destiny (Andersson, 2009) (Debashish Bhattacharya T. N., 2008).

Furthermore, Andersson's work has emphasized that even plastids, such as chloroplasts, can undergo further integration into the host cell during a secondary endosymbiotic event. This integration extends beyond the physical presence of the organelles and delves into the realm of genetic collaboration, where genes from these organelles become part of the host's genetic library (Michelle M Leger, 2017).

2.5 Evolution Beyond Original Functions

As endosymbiotic theory progressed, there was limited understanding of the ability of organelles to evolve within the host beyond their original biochemical functions. Conventional thought at the time was that organelles that showed distinct biochemical properties and mechanisms must have evolved from separate endosymbiotic events. William F. Martin and his team explored the possible

Figure 2-16. William Martin University of Düsseldorf

origins of the hydrogenosome[25], specialized organelles in anaerobic[26] protists and certain fungi. Metabolically anaerobic, hydrogenosomes produce H_2 (hydrogen gas) as a waste product of their metabolic activities. This peculiar feature led to questions regarding their evolutionary origin and potential connection to mitochondria, another critical organelle (William Martin M. M., 1998) (T. Martin Embley, 2006).

Martin's research called into question this prevailing understanding of the origin of hydrogenosomes and mitochondria. It was the widely accepted belief that mitochondria, the energy-producing organelles, originated from the endosymbiosis of Alpha-Proteobacteria[27]. However, using the SSU rRNA technique Carl Woese and James Fox pioneered, Martin uncovered compelling evidence for a phylogenetic relationship between the two organelles (William Martin M. M., 1998).

Martin mapped potential evolutionary pathways by examining the SSU rRNA sequences of both organelles. His findings indicated that the hydrogenosome might represent a modified form of mitochondria, an adaptation stemming from the need to generate energy in oxygen-poor environments. This evolutionary shift from mitochondria resulted in structural and metabolic modifications, leading to the emergence of the hydrogenosome and its production of H_2 as a metabolic byproduct (William Martin M. M., 1998).

---

[25] **Hydrogenosomes**: Energy-producing organelles in some anaerobic eukaryotic microorganisms, similar to mitochondria but generating hydrogen and ATP without oxygen.

[26] **Anaerobic:** Processes or organisms that do not require oxygen for growth and survival, often occurring in oxygen-depleted environments.

[27] **Alpha-proteobacteria**: A diverse group of Gram-negative bacteria that play various roles in nitrogen fixation, photosynthesis, and as precursors of mitochondria in eukaryotes.

Martin's suggested the original mitochondria might have evolved from a precursor organelle, and both mitochondria and hydrogenosomes may share a common ancestry. This dynamic view posited that organelles aren't static but continually evolving, adapting to the host's environmental conditions and needs (T. Martin Embley, 2006).

<u>2.6 Changing Symbiotic Relationships</u>

Figure 2-17 Debashish Bhattacharya Rutgers, The State University of New Jersey

The theoretical and experimental framework built by Margulis, Kessing, Martin, and others opened the door for a deeper understanding of the evolution and origins of algae. With this foundation in place, researchers Debashish Bhattacharya and Hwan Su Yoon set out to study the intricate endosymbiotic relationships in algae and document the convoluted evolutionary relationships among their groups. Their research peered into the vast diversity of plastids within algae, illuminating the range of photosynthetic pigments they possessed. Their findings suggest that endosymbiotic events have led to the emergence of three distinct phylogenetic lines. Primary endosymbiosis involving cyanobacteria and eukaryotic cells resulted in the evolution of green and red algae. Evidence also hints that

Figure 2-18 Hwan Su Yoon Professor, Sungkyunkwan University

dinoflagellates[28] may have evolved through a secondary endosymbiotic event, wherein an unknown eukaryotic cell engulfed another algal species already possessing plastids from a prior endosymbiotic event. (Debashish Bhattacharya H. S., 2003)

Bhattacharya and Yoon's research highlighted that tertiary endosymbiotic events are common in the evolutionary history of dinoflagellates. There may even be cases of dinoflagellates that lost their original plastid only to gain new plastids in a new endosymbiotic event. (Debashish Bhattacharya H. S., 2003)

2.7 Symbiotic Tugs on the Course of Evolution

Figure 2-19 Jessica .M. Nelson Teacher Maastricht Science Programme Faculty of Science and Engineering

While the endosymbiotic theory traditionally connects organelles like plastids and mitochondria to ancient symbiotic events, research by Jessica M. Nelson, Duncan A. Hauser, and Fay-Wei Li shed light on the current symbiotic relationships between free-living organisms such as cyanobacteria and hornworts. Their studies delve into how these nonvascular plants form specialized structures to house cyanobacteria, which fix atmospheric nitrogen, much like root nodules in legumes. This symbiosis benefits the plants by utilizing the cyanobacteria's nitrogen-fixation capabilities (Jessica M. Nelson, 2020).

---

[28] **Dinoflagellates:** A group of single-celled, often photosynthetic algae, notable for their bioluminescence and forming harmful algal blooms in marine and freshwater environments.

The work of Nelson, Hauser, and Li reveals that cyanobacteria within these symbiotic relationships display significant adaptability. They have evolved mechanisms to adjust their nitrogen-fixation [29] rate in response to the environmental conditions and the host plants' needs. These cyanobacteria can also alter their cell walls to meet challenges within the host environment, such as resisting osmotic pressures[30] and toxins (Jessica M. Nelson, 2020).

Figure 2-20 Fay Wei Li Boyce Thompson Institute

Moreover, their studies point to the complex communication systems developed by symbiotic cyanobacteria, enabling efficient signaling of stress conditions and communicating metabolic needs to their host plants. Their research also indicates that symbiotic cyanobacteria have sensing mechanisms that, alongside nitrogen fixation regulation, help balance nitrogen production, preventing potential harm to their host plants (Jessica M. Nelson, 2020). Overall, the findings underscore that endosymbiosis is a continually evolving process, with contemporary examples evident in current plant-cyanobacteria interactions. Their research highlights the influence of endosymbiotic

Figure 2-21 Duncan A. Hauser Boyce Thompson Institute

Figure 2-22 thalloid liverwort (Credit: Jason Hollinger/flickr.com)

---

[29] **Nitrogen fixation:** The process by which atmospheric nitrogen (N_2) is converted into ammonia (NH_3), a form usable by living organisms, by certain bacteria.

[30] **Osmotic pressure**: The force caused by the difference in solute concentration across a semipermeable membrane, driving the movement of water to balance concentrations.

relationships on the evolution of cyanobacteria and algae does not need to end with the integration of the organisms into the biology of their host. The path of algal evolution is also influenced subtly with tiny tugs and nudges as both organisms evolve to sustain the relationship while retaining their biological autonomy.

2.8 Unanswered Questions

Figure 2-23 Edouard Jurkevitch Microbial Ecology Laboratory Hebrew University of Jerusalem

It is essential to recognize the unanswered questions within endosymbiotic theory. A key question revolves around the necessity of phagotrophy[31]—a process observed exclusively in eukaryotic cells—for endosymbiosis. Addressing the origin of the first eukaryotic cells presents a problem similar to the 'chicken and egg' scenario. Yaacov Davidov and Edouard Jurkevitch from the Hebrew University of Jerusalem posited that eukaryotic phagotrophic traits might have evolved from interactions with predatory prokaryotic microbes. Their proposal suggests these predatory microbes could have transitioned from merely penetrating and digesting prey to establishing a more mutualistic, endosymbiotic existence (Yaacov Davidov, 2009).

------

[31] **Phagotrophy**: The process by which a cell engulfs solid particles to form an internal compartment where digestion occurs, typical in many eukaryotes.

The model suggests that a predatory aerobic bacterium invaded the periplasm-like[32] space of an archaeal[33] host, which, over time, evolved into the ancestor of all eukaryotes. This transition facilitated closer contact between the partners, allowing molecular exchanges, gene flow, and the emergence of the eukaryotic cell's complex structure. Davidov and Jurkevitch suggest that the interaction between the two prokaryotes began as a predatory or parasitic relationship that evolved into a symbiotic mutualistic interaction by overcoming conflicts and incompatibilities. This transition is crucial in understanding the evolutionary steps leading to complex eukaryotic cells, where initially independent prokaryotic organisms formed a stable, endosymbiotic relationship through a series of adaptive changes (Yaacov Davidov, 2009).

2.9 In Summary

The development of endosymbiotic theory has been nothing short of revolutionary, threading together disparate strands of biology into a coherent narrative that reshapes our understanding of cellular life's very origins and complexities. What started as mere musings in the mind of Konstantin Merechkowski in the early 20th century, suggesting a symbiotic origin of complex cells found its voice in the insights of Lynn Margulis in the latter half of the same century. Margulis championed that eukaryotic cells, with their compartmentalized structures and sophisticated functions, didn't emerge through incremental adaptations alone but also

---

[32] **Periplasm:** A gel-like space between the outer membrane and the plasma membrane in Gram-negative bacteria, containing enzymes and proteins for substrate binding and degradation.

[33] **Archaea:** A domain of single-celled microorganisms that are prokaryotic, like bacteria, but have unique biochemistry and genetics distinct from bacteria and eukaryotes.

through symbiotic relationships, where life forms cooperative partnerships for mutual survival.

Through their groundbreaking genetic research on the domain of life, Carl Woese and George Fox solidified the understanding that the tree of life isn't merely bifurcated but extends its branches across three domains: Bacteria, Archaea, and Eukarya. Their work, in tandem with the endosymbiosis theory, emphasized the importance of mutualistic interactions in driving evolution, illustrating that collaboration, as much as competition, is at the heart of life's diversity.

As our knowledge deepened, researchers like Wendy Schluchter illuminated the intricate relationships between cyanobacteria and nonvascular plants, showcasing the dynamic nature of endosymbiosis. Schluchter's research reinforced that the symbiotic interplay between organisms isn't relegated to the annals of evolutionary history but is a continuing, dynamic relationship, adapting and evolving to environmental pressures and demands.

So, why is this profound understanding of endosymbiosis significant to biology? It underscores the interdependence of life, highlighting that even at a cellular level, collaboration can lead to innovation and survival. This theory stitches together microbiology, botany, genetics, and evolutionary biology, offering an interdisciplinary lens to comprehend life's intricacies. It challenges the conventional Darwinian narrative by suggesting that cooperation and integration can be as essential to evolutionary advancement as competition and mutation.

As we pivot our focus to the realms of algae and cyanobacteria, the implications of endosymbiotic theory become even more pronounced. Both algae and cyanobacteria have played pivotal roles in the history of Earth's atmosphere, oceans, and terrestrial environments. Their ability to photosynthesize, their interactions with various organisms, and their involvement in numerous endosymbiotic events have made them central characters in the narrative of life on Earth.

Cyanobacteria, with their nitrogen-fixing capabilities, are the linchpins of many ecosystems, enabling life to flourish in diverse habitats. Their symbiotic relationships, as illuminated by Schluchter's work, provide a glimpse into the adaptability and resilience of these microorganisms. Algae, on the other hand, span a vast diversity from microscopic phytoplankton[34] to massive kelp forests. Their symbiotic and competitive interactions have shaped marine and freshwater ecosystems globally.

As we delve deeper into the subsequent chapters, we'll explore the intricate tapestry of relationships between algae, cyanobacteria, and their host organisms. We'll unearth the stories of mutualism, conflict, adaptation, and evolution that have shaped the world as we know it. In understanding endosymbiosis within these contexts, we appreciate life's history and gain insights into potential future trajectories, especially in a world undergoing rapid environmental change.

In conclusion, the history of the development of endosymbiotic theory is a testament to the ever-evolving nature of scientific understanding. This theory, once a fringe idea, has now become a cornerstone of modern biology, offering us profound insights into the interconnectedness of life. The foundational work of Merechkowski, Margulis, Woese, Fox, and Schluchter, among others, paints a picture of life on Earth as a symphony of interactions, where organisms, big and small, constantly shape and are shaped by their relationships.

---

[34] **Phytoplankton:** Microscopic, photosynthetic organisms found in aquatic ecosystems, serving as a primary food source and key producers of oxygen.

The significance of this theory extends beyond the realm of academic curiosity. It reshapes how we view the natural world and our place within it. No organism is truly an island, no matter how advanced or independent it seems. Countless symbiotic relationships, forged, broken, and reformed over the vast expanse of evolutionary time, produced all life on earth.

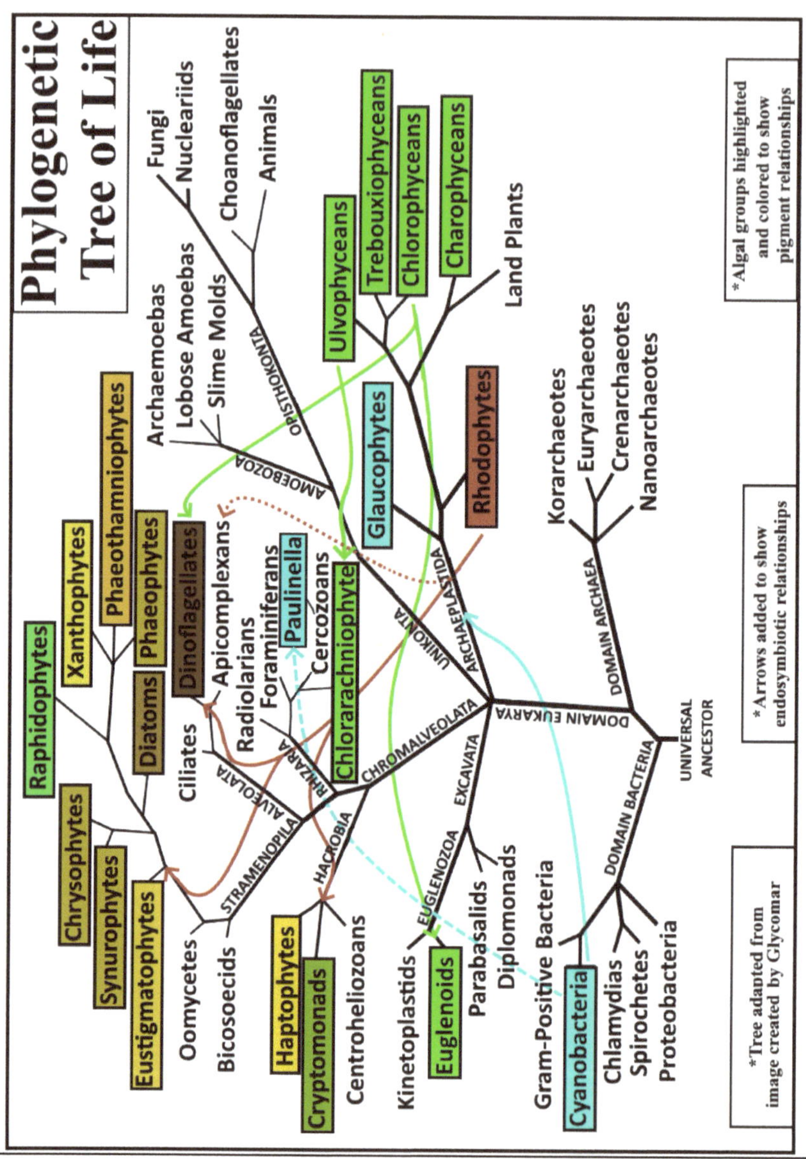

3-1. The above image shows the different lineages of life, including land plants and animals. The highlighted boxes represent groups of organisms that fall under the algal umbrella term. The coloration of the highlighted boxes illustrates the relationships in terms of photosynthetic pigments. This phylogenetic tree also helps to demonstrate that the label algae is a loose term and that algae are not an unbroken lineage of organisms but a broad classification.

3. THE ANCIENT AND MODERN DANCE OF ALGAE AND CYANOBACTERIA

3.1 From Hellscape to Eden

Figure 3-2. Graphic art by Dan Durda/Southwest Research Institute

In an ancient inhospitable world, a hellish landscape spanned the entire Earth, yet this crucible of extremes set the stage for life's emergence. A planet covered with oceans of lava and floating islands of solidified crust. Volcanic eruptions pumped untold tons of toxic and poisonous gasses into the atmosphere. This period was what's known as the Hadean Eon[35] (Rafferty, 2023) (Lunine, 1998). The name is an apt description referencing the Greek god of the underworld, Hades (Lunine, 1998). Even in this hellish alien world, the seeds for life had already been sown.

---

[35] **Hadean Eon**: Earth's earliest geologic eon, spanning from 4.6 to about 4 billion years ago, was characterized by intense heat, volcanic activity, and a molten surface.

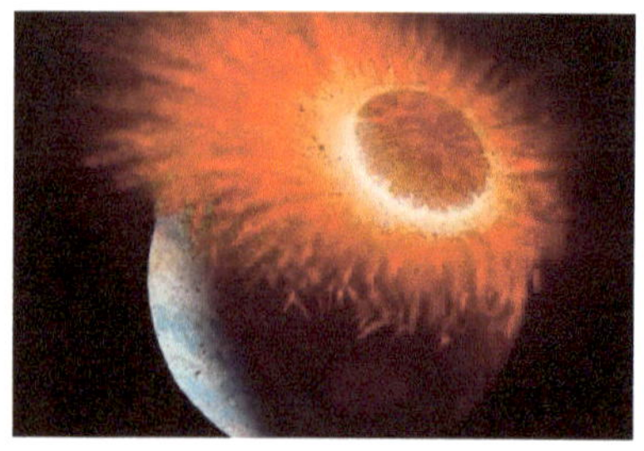

Meteors and comets rained down on early earth, seeding it with amino acids[36], water, and other compounds (Daniel P. Glavin, 2004). A Mars-sized protoplanet collided with early Earth and ejected countless trillions of tons of debris that coalesced into our moon, providing a stable rotation to the earth (W. Benz, 1986). Volcanos spewed compounds such as sulfur and water vapor and provided an outlet for heat trapped within the planet to escape. This process cooled down the Earth enough for a stable crust to form. On this crust, oceans formed, and tides and wind allowed for erosion. This short period of about 15% of earth's history set the stage for one of the most mysterious events—the formation of the first life.

Figure 3-3. Art by William Hartmann, University of Hawai'i at Hilo

The atmosphere of early earth was a toxic mix of nitrogen, ammonia, and sulfur, devoid of oxygen. The traces of the primordial[37] atmosphere still exist in rocks and minerals formed during this early period. Minerals such as uraninite and pyrite, easily oxidizable in oxygen-rich atmospheres, are present in these ancient rocks. Evidence that these rocks formed in an early oxygen-deprived environment. Isotopes of Sulfur and carbon, present in differing ratios, hint at Earth's early atmosphere composition (KASTING, 1993).

Figure 3-4. Under water volcanic vent | Sources/Usage: Public Domain

---

[36] **Amino acids:** Organic compounds that serve as the building blocks of proteins, containing a basic amino group, an acidic carboxyl group, and a unique side chain.
[37] **Primordial:** Refers to the original or earliest state of the Earth's atmosphere.

It is in this seemingly uninhabitable hellscape that life first took hold. Simple, self-replicating molecules [38] most likely emerged near volcanic vents in the deep ocean. An environment rich in nutrients and organic compounds (Hans Kuhn, 1981). A place with extreme cycles of hot and cold. It is a perfect place for the process of life to begin. But this life would seem alien to us today. Simple life, most likely autotrophic[39], given that predatory behavior would most likely not arise in a world void of prey to hunt. The structures of these early life forms were most certainly prokaryotic. The most likely form of metabolism for these pioneering creatures was chemosynthesis[40], oxidizing[41] compounds to produce the energy needed for life's processes. Traces of this primitive ecosystem can still be seen in present-day hydrothermal vents where extremophiles[42] cling to life in the deep, light-starved regions of the ocean (William Martin J. B., 2008).

As life ventured beyond the depths of the primitive oceans of the Archean Eon[43], far away from the energy sources of the hydrothermal vents, life needed a new energy source. This form of energy came in the

---

[38] **Self-replicating Molecules**: During Earth's early pre-life days, self-replicating molecules could copy themselves, precursors to DNA and RNA, essential for the emergence of life from the chemical soup of the ancient world.

[39] **Autotrophic organisms** produce their own food using light, water, carbon dioxide, or other chemicals, without needing to consume other organisms for energy or nutrients.

[40] **Chemosynthesis:** A process by which organisms produce food using chemical energy, typically from reactions involving inorganic molecules, without sunlight.

[41] **Oxidizing:** A chemical reaction where a substance loses electrons, often associated with the gain of oxygen or the loss of hydrogen.

[42] **Extremophiles:** Organisms that thrive in extreme environmental conditions, such as very high or low temperatures, acidity, salinity, or pressure.

[43] **Archean Eon:** A geologic eon 4 to 2.5 billion years ago, characterized by the formation of the Earth's crust and the earliest known life forms.

form of light. Early forms of photosynthesis [44] were most likely anoxygenic[45]. These sulfur-based photosynthesis schemes produced no oxygen and used hydrogen sulfide or elemental forms of sulfur as an electron donor[46]. This type of photosynthesis persists today in some extreme anoxygenic environments and stratified water ecosystems [47] (Olson, 2006).

These early bacteria[48] may have contained a compound known as bacteriochlorophyll[49]. This sulfur-utilizing version of chlorophyll can absorb wavelengths of light[50], including infrared light. Today, bacteria still utilize this method in some extreme environments. Purple and Green sulfur bacteria exist today in the lower depths of microbial mats in select marine niches[51]. When the bacteria exploit hydrogen sulfide, elemental

---

[44] **Photosynthesis**: Process by which plants, algae, and some bacteria convert light energy into chemical energy, producing oxygen and glucose from water and carbon dioxide.

[45] **Anoxygenic**: Pertaining to a process that produces energy through photosynthesis without releasing oxygen as a byproduct, in contrast to oxygenic photosynthesis in plants.

[46] **Electron Donor**: A compound that provides electrons to another in a energy producing reaction, playing a crucial role in energy transfer and chemical reactions.

[47] **Stratified Water Ecosystems:** Aquatic systems with distinct layers that do not mix, each with different conditions like oxygenation, temperature, or salinity, supporting diverse biological communities.

[48] **Bacteria**: Single-celled microorganisms that are ubiquitous on Earth, diverse in form and function, some causing disease while others are essential for ecological balance.

[49] **Bacteriochlorophyll**: Photosynthetic pigments found in various photosynthetic bacteria, allowing them to capture light energy for food production, distinct from chlorophyll in plants.

[50] **Wavelengths of Light**: Distinct measures of light's electromagnetic spectrum, varying in energy and perceived as different colors by the human eye, crucial for various biological processes.

[51] **Niches**: Unique roles or positions of organisms within an ecosystem, involving specific interactions with other biotic and abiotic factors, crucial for maintaining biodiversity.

sulfur may be produced and stored as sulfur grains or granules inside the bacterial cells. This sulfur undergoes oxidation to form sulfate ions[52] (SO_4^{2-}) (Cardona, 2016).

Figure 3-5. Picture of the Roscoff Aber bay affected by green macroalgae deposits (a) with details of a purple sulphur bacterial mat on top of magroalgal deposits (b), and details of three different types of mats selected for the study (HD: high density, MD: medium density, and LD: low density of purple sulphur bacteria) (Cédric Hubas, 2023).

The photosynthetic pigments[53] of many sulfur-utilizing bacteria can exist in a lipid monolayer[54] - often rod-shaped or ellipsoid in appearance. Unlike thylakoid membrane[55] structures in plants, cyanobacteria and algae, these structures are not membrane-bound. These

---

[52] **Ions**: Atoms or molecules with a net electric charge due to the loss or gain of one or more electrons, important in chemical reactions and biological functions.

[53] **Photosynthetic Pigments:** Molecules that absorb specific wavelengths of light and convert it into chemical energy during photosynthesis, such as chlorophyll in plants.

[54] **Lipid Monolayer**: A single layer of lipid molecules with hydrophilic (water-attracting) heads facing water and hydrophobic (water-repelling) tails facing away, found in some cell membranes.

[55] **Thylakoid Membrane**: Membranes within chloroplasts that contain photosynthetic pigments; site of the light-dependent reactions of photosynthesis in plants and algae.

structures, known as chlorosomes[56], are densely packed with well-organized bacteriochlorophyll molecules. Bacteriochlorophyll exists in variants from A-E. Each variant can absorb slightly different wavelengths of light (Chris Greening, 2020).

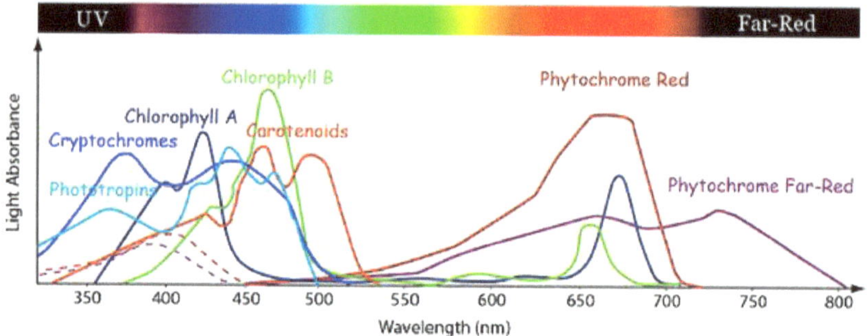

Figure 3-6. Illustration of various photosynthetic pigments and the wavelengths (colors) they absorb

The protein-rich baseplate structure of the chlorosomes facilitates the movement or flow of absorbed energy from the chlorosomes to the reaction centers housed in the cell membrane[57]. The transferred energy drives a series of electron transfer systems[58] (Chris Greening, 2020). Although we may never know the exact structural mechanisms of sulfur-based photosynthesis in the Archean Eon, the contemporary examples of ongoing sulfur photosynthesis provide insight into what these primitive life forms might have been like.

---

[56] **Chlorosomes**: Light-harvesting complexes found in green sulfur bacteria and some green filamentous anoxygenic phototrophs, containing bacteriochlorophyll pigments for capturing low light.

[57] **Cell Membrane**: A biological membrane that separates and protects the interior of all cells from the external environment, facilitating communication and substance exchange.

[58] **Electron Transfer Chain:** A series of compounds that transfer electrons from electron donors to electron acceptors via redox reactions, releasing energy used to produce ATP.

3.2 Cyanobacteria

In this world dominated by bacteria and devoid of oxygen, the first significant phylogenetic step occurred: the rise of cyanobacteria. Named as such, they were once considered blue-green algae due to their blue-green pigments (Malihe Mehdizadeh Allaf, 2022). As bacterial forms expanded

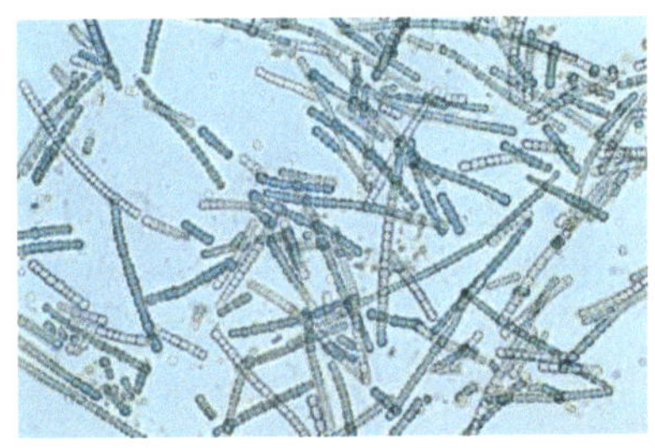

Figure 3-7. A species of blue-green algae, *Cylindrospermum sp*, observed under magnification at the Adelaide laboratories of CSIRO Land and Water in 1993.

throughout the early earth, they evolved to exploit the myriad of available niches. It's believed that cyanobacteria were the first organisms to utilize water as an electron donor for photosynthesis (Patricia Sánchez-Baracaldo, 2022). In a divergence from the traditional anoxygenic photosynthesis and sulfur utilization, cyanobacteria evolved pigment-protein complexes such as Photosystem I[59] and II[60], which allowed for water utilization and oxygen release (Catherine F. Demoulin, 2019).

---

[59] **Photosystem I**: A complex of proteins and pigments in chloroplasts and cyanobacteria that uses light energy to produce the high-energy electron carrier NADPH.

[60] **Photosystem II**: The first protein complex in the light-dependent reactions of photosynthesis and cyanobacteria, which splits water molecules to generate oxygen and electrons.

Figure 3-8. The stromatolites at Shark Bay | Gov. of Western Australia Department of Mines, Insustry Regulation and Safety

The evidence for these events can be observed today. Formations of fossilized microbial mats[61] known as stromatolites [62] can be found throughout the fossil record from about 3.5 billion years ago until about 1 billion years ago (Catherine F. Demoulin, 2019). The disappearance may result from the eventual rise of graze feeders[63] that fed off of these mats. Stromatolites can still be found in isolated regions today; one example is Shark Bay in Western Australia.

Stromatolites are one of the oldest formations in the fossil record. These formations form when cyanobacteria entrap sediments within their microbial mats. The process of cyanobacteria photosynthesis produces a sticky mucus that traps sediment. As the cyanobacteria grow, layers of sediment are laid down, trapped, and hardened. Over time, cyanobacteria evolved into a diverse spread of species, exhibiting the first signs of cooperation and specialization. Some cyanobacteria form filamentous[64]

---

[61] **Microbial Mats**: Layered biofilms, often found in aquatic environments, formed by communities of microorganisms, including bacteria and archaea, which can photosynthesize or oxidize sulfur.

[62] **Stromatolites**: Layered sedimentary formations created by the growth of microbial mats, primarily cyanobacteria, which trap and bind particles, often considered Earth's oldest fossils.

[63] **Graze Feeders**: Organisms that feed on cyanobacteria by grazing on microbial mats and stromatolites, playing a role in shaping these ecosystems through their feeding activity.

[64] **Filamentous**: Descriptive of organisms, like many cyanobacteria, that form long, thread-like chains of cells, which contribute to the structure of microbial mats.

colonies with specialized cells with thick cell walls that act like spores[65] and allow the cyanobacteria to survive when the environment has become less hospitable. Some cyanobacteria developed the ability to control buoyancy and regulate exposure to light using specialized protein structures known as gas vesicles[66]. Yet others developed specialized cells called heterocysts[67], which could convert atmospheric nitrogen into ammonia—a process that will eventually become crucial for many endosymbiotic events (Bettina E. Schirrmeister, 2015).

As the cyanobacterial lineage flourished, these microscopic yet mighty organisms began a silent but profound revolution known as the Great Oxidation Event (GOE) around 2.4 billion years ago(Shih, 2015)(Alcott et al., 2022). This pivotal chapter in Earth's history saw the atmospheric composition undergo a radical shift as oxygen, produced by the photosynthetic activities of cyanobacteria, started to accumulate in the air. It was a gradual process, unfolding over hundreds of millions of years. Still, its impact was monumental, transforming the planet's biochemistry and setting the stage for oxidative weathering processes that sculpted the surface of the Earth(Olejarz et al., n.d.).

GOE's significance extends far beyond the geochemical changes; a biological upheaval redefined the course of evolution. The increased atmospheric oxygen created new ecological niches, leading to evolutionary pressures that would eventually give rise to aerobic

---

[65] **Spores**: Resistant, dormant forms of bacteria and other organisms that can withstand adverse conditions until becoming active again under favorable environmental circumstances.

[66] **Gas Vesicles:** Protein-bound structures in some aquatic bacteria, including cyanobacteria, which provide buoyancy, allowing cells to position themselves optimally for light and nutrients.

[67] **Heterocysts**: Specialized nitrogen-fixing cells found in some filamentous cyanobacteria, providing an anaerobic environment for the fixation of atmospheric nitrogen into ammonia.

respiration—a more efficient way to harness energy than the anaerobic processes that preceded it. As a result, life became more energetic, varied, and complex(Olejarz et al., n.d.). This oxidation event reshaped the biochemistry of the entire planet and set the stage for the story that was about to unfold in evolutionary history. Not bad for tiny prokaryotes.

3.3 The Rise of Green Algae

Although the exact moment in geologic history is unclear, sometime during the latter half of the Precambrian[68] (Endosymbiosis and the Evolution of Eukaryotes, n.d.), the first significant symbiotic event occurred with profound implications for life on Earth. An ancient predatory heterotrophic cell[69], which is still unknown to science (William F. Martin, 2015), Consumed a species of cyanobacteria. However, unlike most meals consumed by this cell, this particular meal did not digest. Instead, this cyanobacteria survived within this cell as a prisoner. This chance occurrence, deep in geologic time, caused the dynamics of the microbial world to change forever. A protective vesicle may have formed around the cyanobacteria, allowing it to survive and even reproduce within the cell. All the while, the cyanobacteria produced excess nutrients for its host cell through its photosynthetic capabilities. Over time and countless successive generations, the host species, having a readily available energy source within its own body, may have grown dependent. Subsequent generations evolved to protect and nurture the intruder to reap this unlikely

---

[68] **Precambrian**: Geological time period before Cambrian, spanning Earth's formation to complex life emergence, approximately 4.6 billion to 541 million years ago.
[69] **Heterotrophic cell**: A cell that cannot synthesize its own food and relies on intake of organic substances for nutrition and energy.

relationship's benefits (Endosymbiosis and the Evolution of Eukaryotes, n.d.).

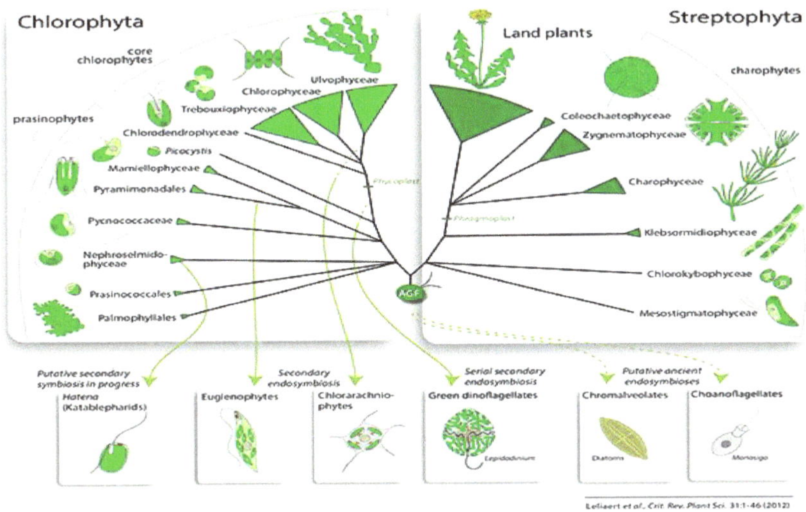

Figure 3-9. Summary of the green lineage's evolutionary tree (top) and the distribution of green genetic traits among other eukaryotes (bottom). (Frederik Leliaert, 2012)

As often happens in evolutionary history, traits, and processes which were no longer needed tended to filter out through natural selection[70]. The host cell may have lost its heterotrophic nature, while the cyanobacteria symbiont started to lose its independent motility[71] and other processes (Oborník, 2019). The cyanobacteria slowly evolved into photosynthetic organelles, losing their individuality and becoming almost indistinguishable from the different parts and functions within the host cell. It is most likely this process that gave rise to early algae (I. N. Stadnichuk, 2021). Genetic traces of cyanobacteria can be found in

---

[70] **Natural selection**: Evolutionary process where organisms better adapted to their environment tend to survive and produce more offspring.

[71] **Motility**: The ability of an organism or cell to move independently, using metabolic energy.

modern chloroplast, showing a common ancestor between them (U. Kutschera, 2005).

Figure 3-10.
Taxonomical, morphological and ecological diversity among green algae.
A: Pterosperma   B: Nephroselmis   C: Palmophyllum D: Tetraselmis
E: Chlorella   F: Oocystis   G: HaematococcusH: Pediastrum
I: Bulbochaete   J: Chaetophora   K: Ulothrix   L: Ulva
M: Cladophora   N: Boergesenia   O: Acetabularia P: Caulerpa
Q: KlebsormidiumR: Spirotaenia   S: Nitella   T: Micrasterias
U: Coleochaete
(Frederik Leliaert, 2012)

The phylogenetic tree of algae has grown with tangled branches, splitting off with each endosymbiotic event. The initial primary endosymbiotic event that led to the eventual evolution of chloroplast, roughly 1.0-1.5 billion years ago, was one in a long line of similar events (Oborník, 2019). This initial event resulted in the evolution of

Chlorophyta[72] and *Charophyta*[73] (Ling Fang, 2017). The *Chlorophyta* and *Charophyta* are divisions of green algae that share many biological characteristics with land plants. They produce and utilize chlorophyll a and b[74] and store starch[75] within plastids (Qing Tang, 2020) (Sarah E. Glass, 2023). It is believed that "true plants[76]" trace their evolutionary origins to a common ancestor of these groups, with *Charophyta* being more closely related to land plants (Ling Fang, 2017). These divisions can be quite varied in morphological[77] characteristics. The structure and morphology of *Charophyta* are more similar to land plants, and it is easy to mistake Charophyta algae, such as *Chara*, for plants due to their morphological similarities (Peter Civáň, 2014).

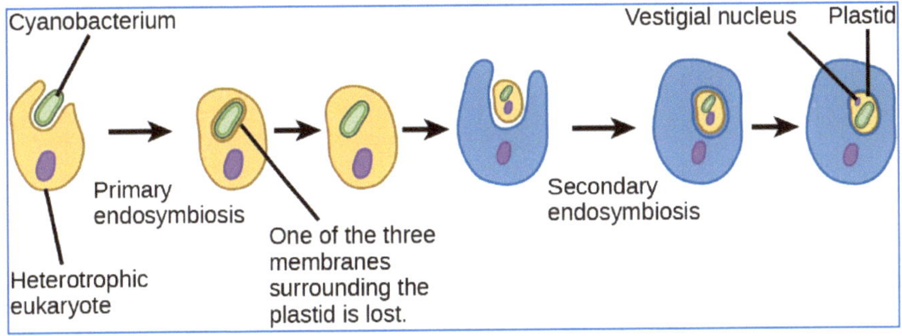

Figure 3-11. Illustration depicting primary and secondary endosymbiotic event between cyanobacteria and unknown heterotrophic organisms (OpenEd CUNY)

---

[72] Chlorophyta: A division of green algae, including various species, primarily aquatic, photosynthetic, with chlorophyll a and b.

[73] **Charophyta**: A division of green algae, often considered the closest relatives of land plants, characterized by complex cells and growth patterns.

[74] **Chlorophyll a and b**: Primary photosynthetic pigments in plants and algae, absorbing light for energy conversion.

[75] **Starch**: A carbohydrate polymer stored by plants as an energy reserve; insoluble in water.

[76] **True plants**: Organisms belonging to the kingdom Plantae, distinct from algae, typically having roots, stems, leaves, and vascular tissue.

[77] **Morphological**: Pertaining to the form and structure of organisms, often used in taxonomy and evolutionary biology.

Chlorophyta comprises a wide range of species demonstrating significant diversity in form and function. Among these are unicellular[78] species such as *Chlamydomonas*, characterized by its flagellated[79] form enabling motility (Xi Li, 2021). *Chlamydomonas* represents a fundamental unicellular model within the *Chlorophyta*, beneficial for studying the fundamental issues related to the transition to multicellularity due to its simplicity and well-studied genetics. Additionally, the *Chlorophyta* includes colony-forming algae[80] like *Volvox*, which are spherical and showcase the evolutionary transition from single-celled to multicellular organisms within the *Volvocine* line of algae (Ling Fang, 2017). This division exhibits a broad spectrum of organizational complexity, ranging from individual unicellular forms to complex colonial structures, highlighting the evolutionary flexibility and adaptability of the *Chlorophyta* (Bennici, 2008).

Specifically, the nonmotile coccoid and colonial green algae within the divisions *Chlorophyta* and *Streptophyta*, such as the freshwater taxa[81] including *Chlamydomonas*, demonstrate the morphological and ecological versatility of the *Chlorophyta*, which can range from unicellular to multicellular and marine taxa like *Ulva* (Peter Civáň, 2014). This diversity is expressed in their morphological structures and ability to inhabit various environments, from freshwater to marine and terrestrial habitats, indicating a broad adaptive capacity that allows multiple independent transitions between these habitats (Xi Li, 2021).

---

[78] **Unicellular**: Comprising a single cell, which performs all necessary functions for life.

[79] **Flagellated**: Having one or more flagella, whip-like appendages enabling mobility in fluid environments.

[80] **Colony-forming algae**: Algae species that group together, forming a colony, exhibiting some level of cooperation.

[81] **Taxa**: Groups of one or more organisms, classified by scientists to organize and understand biological diversity.

3.4 Red Algae as Host and Symbiont

Red algae or *Rhodophyta* can trace their origins to the same endosymbiotic event as the green algae mentioned before (Saunders GW, 2004). Red algae draw their coloration from phycoerythrin[82] and phycocyanin[83] (Michael Borg, 2023).

Figure 3-12. Red Algae Bleaching Coral
Creative Commons Johnmartindavies

These pigments aid in concealing the green of the chlorophyll. Cyanobacteria also contain similar pigments, which red algae likely inherited during the primary endosymbiotic event (Shunsuke Hirookaa, 2022). The pigments phycocyanin and phycoerythrin enhance the efficiency of photosynthesis and protect the red algae from intense light. Red algae can photosynthesize at greater depths than their green counterparts due to these pigments and their ability to absorb blue and green wavelengths of light. These wavelengths can penetrate deeper into water than the red and orange wavelengths utilized by chlorophyll alone (Sofie E. Voerman, 2022).

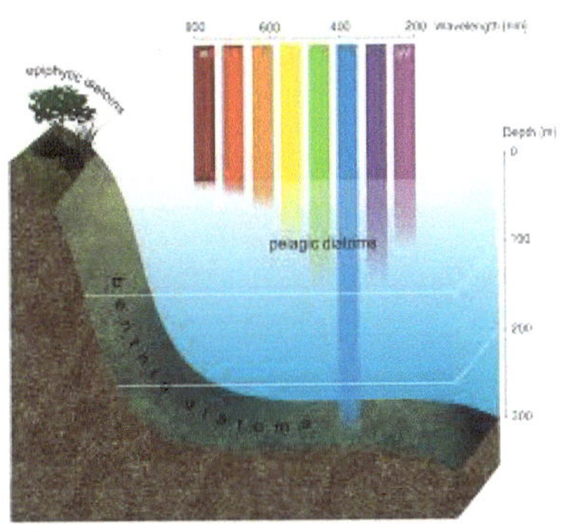

Figure 3-13. Spectra composition of light depending on the depth. (Paulina Kuczynska, 2015)

---

[82] **Phycoerythrin**: A red photosynthetic pigment found in red algae that absorbs blue light and imparts a red color to the algae.

[83] **Phycocyanin**: A blue photosynthetic pigment found in cyanobacteria and some algae, it absorbs red light and aids in photosynthesis.

Although some single-celled red algae exist, most red algae are multicellular, and unlike other algal taxa, red algae lack any flagellated cells and do not possess flagella (Michael Borg, 2023). Although endosymbiosis played a crucial role in the evolution of red algae, the endosymbiotic story did not end for red algae with the primary event. The phylogenetic tree of life of algae branched again, with red algae becoming an unsuspecting symbiont. Between 600 and 800 million years ago, an unknown eukaryotic cell engulfed a red algae. This event, being a secondary endosymbiotic event, resulted in the red algae becoming entrapped in a vacuole, which later evolved into what's called a secondary plastid (Fangru Nan, 2017). Unlike primary plastids in green algae and red algae, which typically contain two membranes, secondary plastids may include three or four membranes due to the secondary endosymbiotic nature of their acquisition. As time and generations passed, much of the red algae's mechanisms were either lost or integrated into the host cell, paving the way for the emergence of diverse algal and protist lineages. These groups include the *Cryptophytes*, which interestingly have a small remnant nucleus called a nucleomorph[84] within their red algae plastids. Cryptophytes have plastids that contain four membranes (Jong Im Kim, 2022).

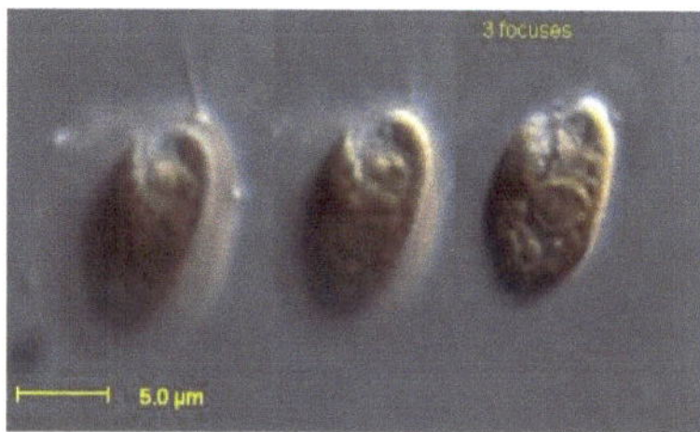

Figure 3-14. *Rhodomonas salina* a member of *Cryptophytes*
By Daniel Vaulot, CNRS, Station Biologique de Roscoff - Own work, CC BY-SA 2.5, https://commons.wikimedia.org/w/index.php?curid=1267713

---

[84] **Nucleomorph**: A remnant nucleus of the endosymbiont within some eukaryotic cells, specifically in some algae, representing a vestige of endosymbiosis.

Not all species within the *Cryptomonas* genus are photosynthetic, providing evidence of the complex path of algal evolution. Non-photosynthetic strains of *Cryptomonas* lost essential photosynthetic genes, possessing only remnants of their plastid genomes, such as pseudogenes [85] related to photosystem proteins (Goro Tanifuji, 2020). These genetic vestiges, along with a range of structural variants and genomic rearrangements, suggest an ongoing process of genome reduction and metabolic adjustment after the loss of photosynthesis (Shigekatsu Suzuki R. M., 2022).

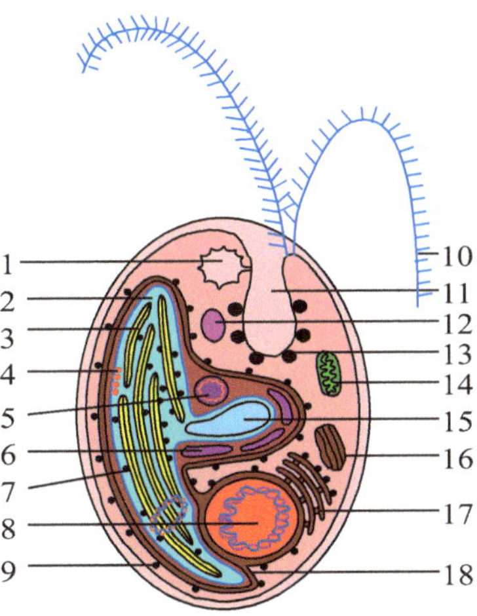

Figure 3-15. This diagram is of the *Cryptophyta* cell scheme. #5 is the nucleomorph remaining from the red algae.
By Franciscosp2 - Own work, CC BY-SA 4.0,

At the same time, some lineages, like *Goniomonas avonlea*, diverge from this pattern, lacking a plastid altogether. The absence of a plastid in *Goniomonas avonlea*, supported by comprehensive genomic and transcriptomic [86] data, suggests non-photosynthetic ancestry rather than secondary plastid loss. Specific metabolic pathways like those for alpha glucan storage polysaccharides, which do not rely on plastid function, further highlight the varied outcomes of endosymbiotic events across different lineages (Ugo Cenci, 2018). The continuous history of *Cryptophytes* suggests tasks necessary or beneficial when plastids are

---

[85] **Pseudogenes**: Non-functional sequences in a genome resembling genes, having lost their gene expression ability or protein-coding potential due to mutations.
[86] **Transcriptomic**: The study of the transcriptome, encompassing all RNA molecules, to understand gene expression patterns within a cell or organism.

present in an organism may continue to function and stay present after the organism loses its plastid over evolutionary time.

Haptophytes, a diverse clade of unicellular algae, including species like *Emiliania huxleyi*, are descendants of red algae endosymbiotic encounters and play a pivotal role in marine ecosystems. They significantly contribute to marine phytoplankton populations, forming substantial algal blooms and contributing considerably to the photosynthetic biomass in the world's oceans (Hisashi Endo, 2018) . *Haptophytes* produce calcified scales, which significantly affect the global carbon cycle. Species such as *Emiliania huxleyi* fix carbon through photosynthesis and calcification, thereby playing a crucial role in carbon sequestration and regulating atmospheric carbon dioxide levels (El Mahdi Bendif, 2023).

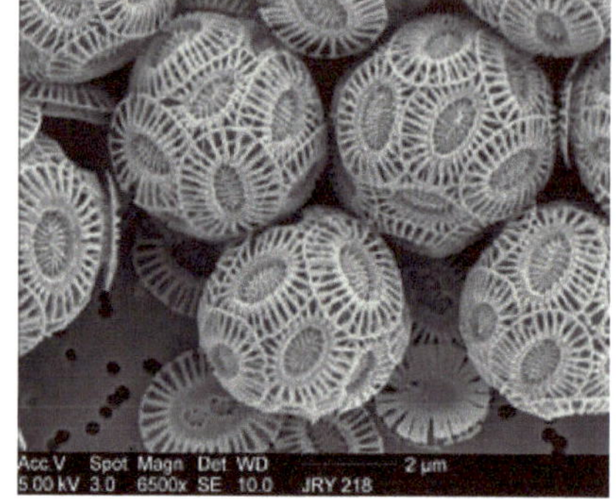

Figure 3-16.
Photo: *Emiliania huxleyi*
Dr. Jeremy R. Young Paleontology Dept.
The Natural History Museum
LONDON, SW7 5BD, UK

3.5 Heterokonts (Stramenopila)

Heterokonts or *Stramenopila* are a diverse phylum[87] of algae and other organisms, including brown algae such as kelp, golden algae, and diatoms. The algal species in this phylum can trace their origins to a past endosymbiotic encounter with red algae (Shenglan Li, 2005). Its important

---

[87] **Phylum**: A taxonomic rank below Kingdom and above Class, grouping organisms based on general body plan and lineage.

to note that the term *"heterokont"* has historically been applied to various algal taxa, creating ambiguity without context. Initially introduced in 1899 to classify certain algae, its meaning has shifted over time, encompassing unrelated groups and losing clarity in scientific discussions (Blackwell, 2009).

"*Heterokont*" comes from the Greek words "heteros," meaning different, and "kontos," meaning pole, due to the arrangement of their two flagella - one flagellum is smooth. At the same time, the other flagella contain hair-like projections called mastigonemes or tripartite hairs (Cavalier-Smith, 2018). In the phylum of *Stramenopiles*, flagella are typically present only in specific reproductive cells. For instance, in many brown algae species belonging to this phylum, the male gametes (sperm cells) are flagellated and utilize these

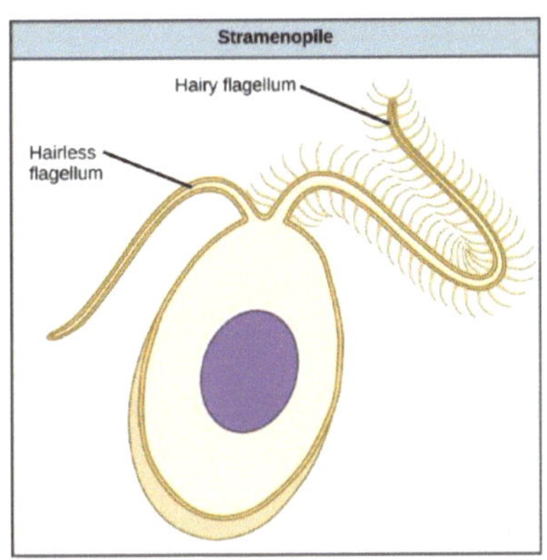

Figure 3-17. Credit: LibreTexts Biology

flagella for motility, akin to the flagella of human sperm cells (Gang Fu, 2014). However, most life stages in multicellular algae, including kelp's large, visible structures, do not display flagella.

The brown algae, or *Phaeophyceae*, are a diverse class of algae characterized by their distinctive brown to olive-green coloration, which results from the dominance of the carotenoid pigment fucoxanthin over chlorophyll (Nur Akmal Solehah Din, 2022). This pigment provides the characteristic color and has a wide photon absorption capacity, which is beneficial for photosynthesis in the low-light marine environments where these algae typically reside (Ulrike Neumann, 2019). The brown algae encompass approximately 1500 to 2000 described species, found chiefly in marine environments, particularly in cold waters along continental coasts (Romain Derelle, 2016). They have evolved complex multicellularity similar to that seen in plants, fungi, and animals. The size and shape of these algae can vary greatly, ranging from microscopic forms to large, complex structures like those seen in kelp forests.

Kelps and seaweeds within this class are differentiated from other members of the phylum by their complex multicellular organization. Kelps, for example, can form massive underwater forests and are some of the largest algae. At the same time, seaweeds are a broader category that includes many different species of marine macroalgae, not all of which belong to the brown algae class. Kelps are generally larger and more

Figure 3-18. Kelp Forest | Douglas Klug / Getty Images

structurally complex than many of the simpler, often smaller seaweeds, though both are important ecological components of their environments (François Thomas, 2014).

The pigment fucoxanthin is particularly advantageous because it absorbs light in the blue-green spectrum, which penetrates deeper into the water than light of longer wavelengths, allowing brown algae to photosynthesize efficiently in deeper, darker waters where other photosynthetic pigments are less effective (Sarah Méresse, 2020). Moreover, fucoxanthin has drawn significant scientific and industrial

Figure 3-19. Credit: Atlantic Sea Farms

interest due to its functional activities in human health, including antioxidant properties. Studies suggest that fucoxanthin could have a variety of applications, from food to pharmaceuticals (Ulrike Neumann, 2019).

Gold algae, also known as golden-brown algae, belong to the class *Chrysophyceae* within the division *Chromophyta*, residing in marine and freshwater environments. As with brown algae, they also contain the pigment fucoxanthin, which imparts a yellow-brown to golden color to these organisms (Dales, 1960). Most golden algae are unicellular biflagellates, meaning they typically have two flagella specialized for movement. They are pretty diverse, with about 33 genera and approximately 1,200 species known (Encyclopaedia, 2023).

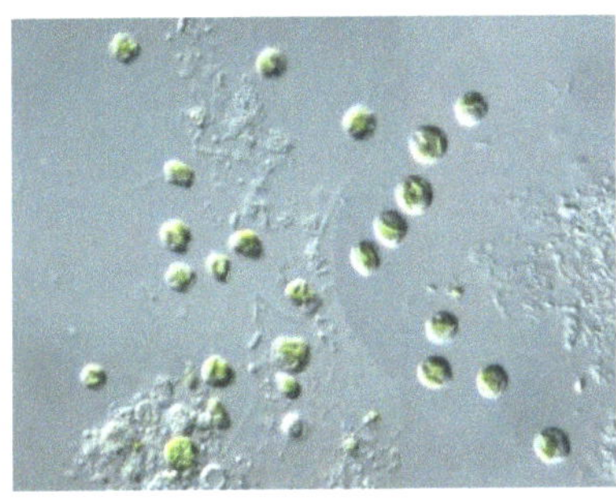

Figure 3-20. *Ochromonas sp.* **2018 Protist Information Server**

Ochromonas, a genus within the golden algae, is particularly notable for its mixotrophic nature. Mixotrophy is the capability of an organism to use both autotrophy (such as photosynthesis) and heterotrophy (such as ingestion of food particles) for energy and carbon acquisition. *Ochromonas* species thrive in various environments, from marine, brackish[88], and freshwater systems and even in extreme environments such as hypersaline[89] ponds and acidic lakes. They can be significant consumers of bacteria but also employ photosynthesis for survival when prey is scarce. This mixotrophic ability is an adaptation that allows these organisms to thrive in various nutrient-limited conditions or where light requirements vary (Alle A. Y. Lie Z. L., 2018).

---

[88] **Brackish**: Water that has more salinity than fresh water, but less than sea water, typically found where rivers meet the sea.

[89] **Hypersaline**: Waters that are extremely salty, often more so than seawater, typically found in areas with high evaporation rates.

The mixotrophic nature of *Ochromonas* species provides insight into the evolution of eukaryotic cells. The ability of *Ochromonas* to switch between feeding methods depending on environmental conditions demonstrates a flexible metabolic system that could have been advantageous during the early evolution of eukaryotic cells. A Genomic study found that while light doesn't significantly impact the population size of *Ochromonas* species, it does stimulate the activation of genes involved in both photosynthesis and the consumption of food particles (phagotrophy). Additionally, when bacteria are available as food, there's a substantial alteration in the pathways that process carbon and nitrogen. The results of this study indicate that the gene activity and metabolic functions of *Ochromonas* are finely tuned to the availability of resources in their environment, hinting at a transitional phase between phagotrophic predators and strictly autotrophic algae (Alle A. Y. Lie Z. L., 2017).

Figure 3-21. Marine diatoms image by Mogana Das Murtey and Patchamuthu Ramasamy

The endosymbiotic story of the diatoms is similar to that of the other algal lineages that trace their origins to the initial red algae endosymbiotic event, tracing back 250 million years. The distinguishing feature of diatoms is their unique morphology. Their common name derives from the Greek prefix "di," or two, and "atomos," also of Greek origin, meaning indivisible or not cut. Their unique unicellular morphology has a notable two-part silica-based[90] shell called a frustule. The frustule, however, is not symmetrical, with the epitheca being slightly larger than the hypotheca and fitting over the hypotheca like the lid of a shoebox. Like other algae

---

[90] **Silica**: A hard, unreactive, colorless compound that occurs as the mineral quartz and as a principal constituent of sand and rocks.

in the red algae endosymbiotic lineage, diatoms possess four membrane secondary plastids (Jürgen F. H. Strassert, 2021).

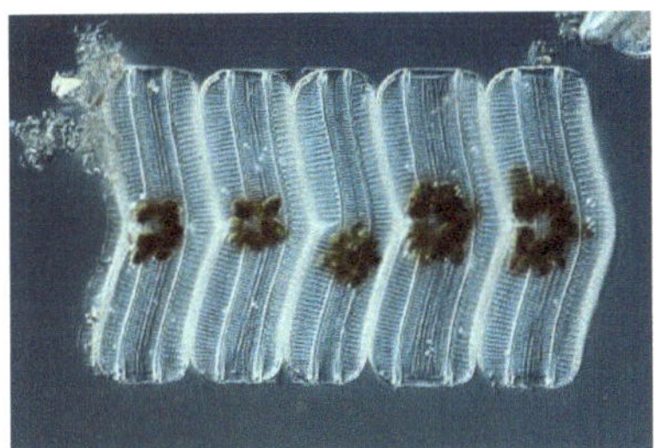

Figure 3-22. *Achnanthes longipes* (a diatom, *Bacillariophyta*) **Frank Fox Image of Distinction 2012 Photomicrography Competition**

Although extremely rare and sparsely documented, some diatoms exhibit mixotrophic behavior, including practicing phagotrophy or engulfing and consuming microbes, namely bacteria. Diatoms such as *Gyrosigma Sp., Achnanthes longipes,* and *Nitzschia alba* are known to have mixotrophic abilities; however, it is unclear whether this practice is a vestigial[91] trait from the early eukaryotic pre-endosymbiotic period or an evolutionary re-emergence due to environmental pressures such as limited access to light and nutrients (Valeria Villanova, 2021).

3.6 Green Algae Symbionts

Around 600 to 800 million years ago, another set of endosymbiotic events involved a secondary endosymbiosis with green algae. These events gave rise to several lineages of algae (Jürgen F. H. Strassert, 2021). As with red algae endosymbiotic events, encounters with green algae were not singular.

Euglenoids, primarily freshwater algae characterized by their flagellar motility, possess plastids resulting from secondary

---

[91] **Vestigial**: An organ or part of an organism that has become reduced or functionless over the course of evolution.

endosymbiosis with a green alga related to the genus *Pyramimonas*. This endosymbiotic event is significant because it provided the *Euglenoids* with plastids containing chlorophyll a and b, which indicates their green algal origins. This endosymbiosis has led to a triple-membraned plastid within the Euglenoids, distinguishing them from other algae with different endosymbiotic histories (Bożena Zakryś 1, 2017).

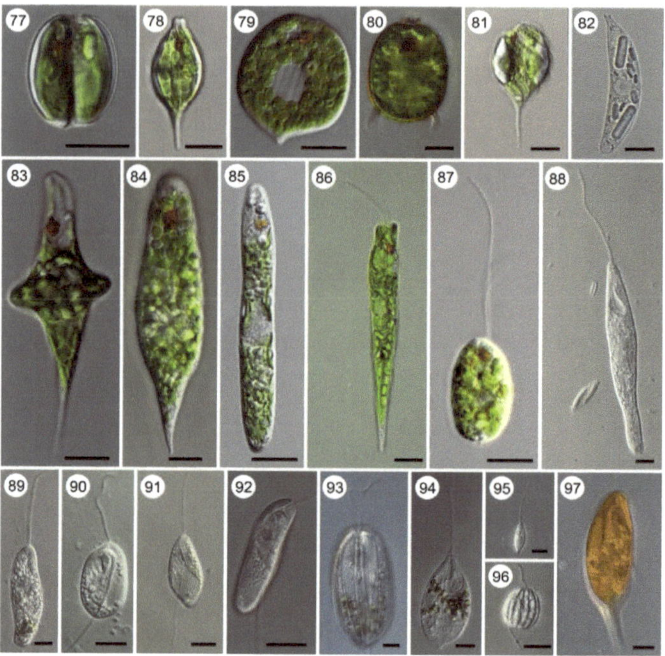

Figure 3-23 This image shows several Euglenid species. Not all species of Euglenid are photosynthetic.

Photosynthetic Euglenids:
(77) *Cryptoglena sp.*   (78) *Lepocinclis autumnalis*
(79) *Phacus acuminatus*   (80) *Trachelomonas armata*
(81) *Monomorphina sp.*   (84) *Euglenaformis proxima*
(85) *Euglena gracilis*   (86) *Eutreptiella pomquetensis*
(87) *Rapaza viridis* (also mixotrophic)

Non-photosynthetic or Unknown:
(82) *Menoidium sp.*   (83) *Discoplastis spathirhyncha*
(88) *Jenningsia sp.*   (89) *Peranema sp.*
(90) *Anisonema sp.*   (91) *Heteronema vittatum*
(92) *Dinema sp.*   (93) *Olkasia polycarbonate*
(94) *Notosolenus ostium*   (95) *Sphenomonas teres*
(96) *Lentomonas corrugate*   (97) *Calkinsia aureus*
(Alexei Y. Kostygov, 2021)

54

The photosynthetic Euglenoids are especially notable for their diverse evolutionary lineages, identified through molecular and genomic studies. These studies suggest that the plastids of *Euglenoids* likely originated from *Prasinophyte* green algae, among the most ancient and basal lineages of green algae (Monique Turmel, 2009) (Shinichiro Maruyama, 2011). Interestingly, the plastid genome of the well-studied Euglenid *Euglena gracilis* shows a high number of introns or non-coding sequences of DNA, which is uncharacteristic when compared to its *prasinophyte* counterparts, indicating a complex evolutionary adaptation following the endosymbiotic event (Jean-François Pombert, 2012).

Some Euglenoids, such as *Rapaza viridis,* exhibit mixotrophy, wherein they engage in phagotrophy, absorbing nutrients by ingesting bacteria and other protists and performing photosynthesis. This mixotrophic capability might be an evolutionary advantage that allows Euglenoids to survive in various environmental conditions, further underscoring the complexity of their adaptation and the multifaceted nature of their endosymbiotic origins (Bożena Zakryś 1, 2017).

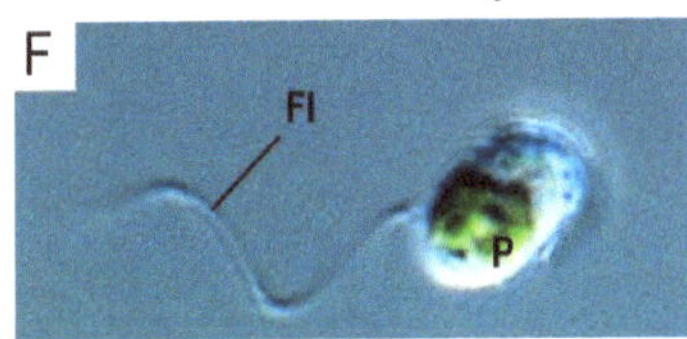

Figure 3-24. Flagellate cell of Lotharella globosa (Chlorarachniophyte) with a single flagellum (Fl) and plastid (P) By Yoshihisa Hirakawa et. al.

Chlorarachniophytes are rare marine algae that derive their secondary plastids from incorporating the green alga *Ulvophyte* (Rafael I Ponce-Toledo, 2018). This difference in source symbiont compared to the Euglenoids highlights that the events are not singular and that these occurrences can happen several times in the history of a phylogenetic lineage (Christopher Jackson, 2018). An intriguing trait of *Chlorarachniophytes* is that their plastids contain a nucleomorph. The presence of a nucleomorph, just as in some plastids descendent of red algae, shows evidence of the secondary endosymbiotic event (Shigekatsu Suzuki Y. H.-I., 2016).

3.7 Dinoflagellates

Dinoflagellates are a superclass of algae that requires its own section within this chapter. Theirs is a story of secondary and tertiary endosymbiosis, serial endosymbiosis, and loss and regaining of traits across time. Many dinoflagellate species have chloroplasts originating from different evolutionary lines, indicating that they did not arise directly from a single endosymbiotic event. Instead, their photosynthetic abilities are often due to secondary endosymbiosis, where they have acquired chloroplasts from organisms like red algae, or tertiary endosymbiosis, involving further complexity in their symbiotic relationships (Debashish Bhattacharya H. S., 2003) (Levandowsky, 2012). Consequently, many dinoflagellate lineages have lost the original primary plastid, if ever they had it, and instead have gained new types of plastids through these later endosymbiotic events (Levandowsky, 2012).

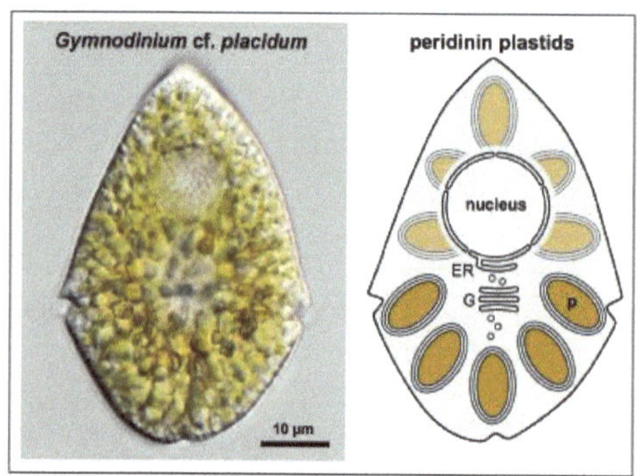

Figure 3-25. (Ross F Waller, 2017)

Additionally, the peridinin-type plastids, a defining feature of dinoflagellates, showcase a secondary endosymbiotic origin. These plastids are characterized by the presence of the carotenoid peridinin, which contributes to their unique light-harvesting capabilities. The peridinin plastid's genome structure is highly specialized, consisting of minicircles with many genes transferred to the host nucleus. Despite the prevalence of peridinin plastids among dinoflagellates, the variation in plastid types within the group points to a complex history of plastid acquisition and evolution, including multiple endosymbiotic events beyond the initial acquisition of the peridinin plastid (Ross F Waller, 2017).

Some dinoflagellates contain plastids of green algal origin, such as *Lepidodinium chlorophorum*, while others, such as *Amphidinium carterae* and *Symbiodinium*, contain red algal-sourced plastids. The green-colored plastids in *Lepidodinium* originated from an

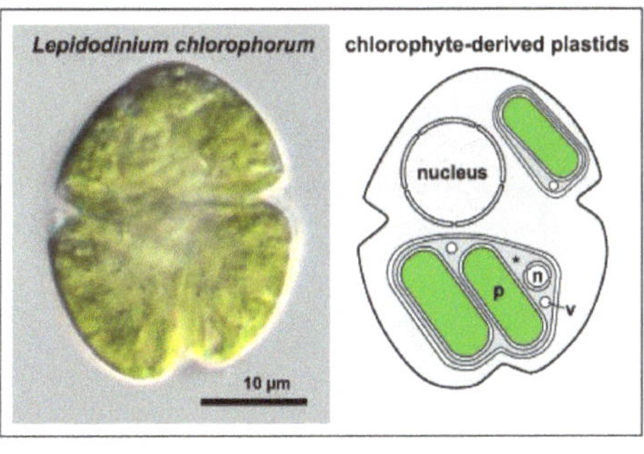

Figure 3-26. (Ross F Waller, 2017)

endosymbiotic *pedinophyte* or a green alga closely related to *pedinophytes* (Ryoma Kamikawa, 2015) (Jeffrey D Leblond, 2012). These plastids are likely due to secondary endosymbiotic events at different points along the dinoflagellate's phylogenetic history. Researchers have confirmed the red algal origin of *Symbiodinium* plastids, which share a common ancestry with the plastids of other alveolates such as apicomplexans (Hwan Su Yoon, 2002) (Jan Janouskovec, 2010).

The relationship between apicomplexans and dinoflagellates presents an intriguing aspect of evolutionary biology, particularly in their shared lineage of plastids. Both groups are part of the broader *Alveolata* supergroup and have plastids believed to be of red algal origin, suggesting a shared evolutionary event. The nonphotosynthetic plastid found in apicomplexan parasites, such as those causing malaria, has sparked a lively debate regarding plastid origin, with molecular data supporting the hypothesis that it shares a red algal heritage with the plastids of related dinoflagellates (Jan Janouskovec, 2010). Apicomplexans have adapted these plastids to serve essential, non-photosynthetic functions crucial for their survival. A later chapter will explore in further depth the fascinating convergence and divergence in the evolution of plastids in these two groups and the resulting functional diversification in apicomplexans.

Dinoflagellate evolution did not stop at secondary endosymbiosis. *Kryptoperidinium foliaceum* is a dinoflagellate with an intriguing story of

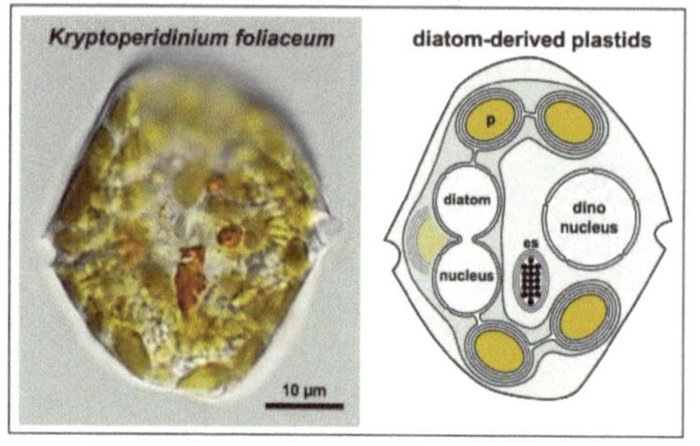

Figure 3-27. (Ross F Waller, 2017)

tertiary endosymbiosis, which may serve as an example of an intermediate stage between initial events and eventual evolution into plastids. *K. Foliaceum* actively hosts diatoms within its cytoplasm [92]. These are not plastids originating from diatoms but biologically separate endosymbiotic diatoms with their own life cycle (Rosa Isabel Figueroa, 2009). The way the diatoms are isolated from the host's biology is intriguing. The diatoms reside within one membrane inside *K. foliaceum*. This unique situation may have resulted from the unique nature of the endosymbiotic event. The single membrane configuration may facilitate more efficient nutrient and signaling molecule exchange (Behzad Imanian P. J., 2007).

An interesting feature of this endosymbiotic relationship is both organisms' full retention of plastids and other organelles. The symbiotic diatom retained all the functionality of its organelles, including the nucleus and photosynthetic plastid. Meanwhile, *K. foliaceum* retained its plastids, which are still capable of photosynthesis (Elisabeth Hehenberger, 2014) (Rosa Isabel Figueroa, 2009). This feature allows the organism to possess multiple sources of metabolism, a situation with a wide range of benefits. Since the dinoflagellate plastid pigment profile differs from diatoms, this may widen the wavelength range for which the entire organism can photosynthesize. Allowing for more energy production and a broader range of niche ecosystems the partnership can exploit, including different light conditions (Behzad Imanian J.-F. P., 2012) (Elisabeth Hehenberger,

---

[92] **Cytoplasm**: The jelly-like substance within a cell, holding organelles and providing a medium for biochemical processes.

2014). It is also possible that nutrient exchange may occur in both directions, allowing for a more efficient metabolic process by both (Behzad Imanian P. J., 2007).

Regarding the reproduction of *K. foliaceum*, little information is available on how the host dinoflagellate retains the endosymbiotic diatoms within its daughter cells. Life history and cell cycle studies show that *K. foliaceum* has a sexual cycle. In the studied strains, the binucleate condition is permanent, although it is unknown how the diatom nucleus behaves during these processes (Rosa Isabel Figueroa, 2009). Observations made of other endosymbiotic relationships can provide insight into this process. For instance, the endosymbiotic relationship of *Paramecium bursaria* and *Chlorella spp.* shows a synchronous life cycle between the endosymbionts and the host, ensuring even symbiont populations in the resulting daughter *P. bursaria* cells. There are several ways *P. bursaria* may control the population of *Chlorella*. Chemical signals may synchronize the host and symbiont cell life cycles to ensure proper reproductive timing (Takahashi, 2016). It is also possible that *P. bursaria* may have a feedback mechanism that can sense the

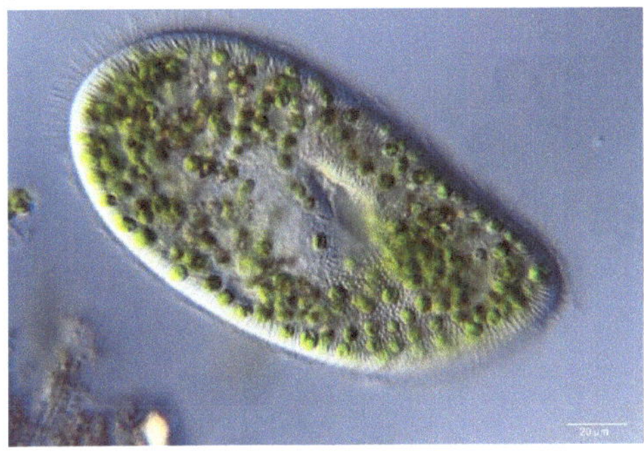

Figure 3-28. *Paramecium bursaria* with green zoochlorellae living inside endosymbiotically |by Anatoly Mikhaltsov

number of *Chlorella* in its cytoplasm, allowing *P. bursaria* to trigger cell division once the *Chlorella* population reaches an unmanageable threshold (Yuuki Kodama, 2022) (Yutaka Kato, 2009). Although the relationship between *P. bursaria* and *Chlorella* provides insight into possible reproductive methods, more research is needed to understand how *K. foliaceum* coordinates reproduction.

Some dinoflagellates, specifically *Karlodinium veneficum* and *Durinskia baltica*, have plastids from a tertiary endosymbiotic event with cryptophyte algae. Cryptophyte algae, in turn, got their plastids from red algae and kept a nucleomorph, a remnant of the red algal

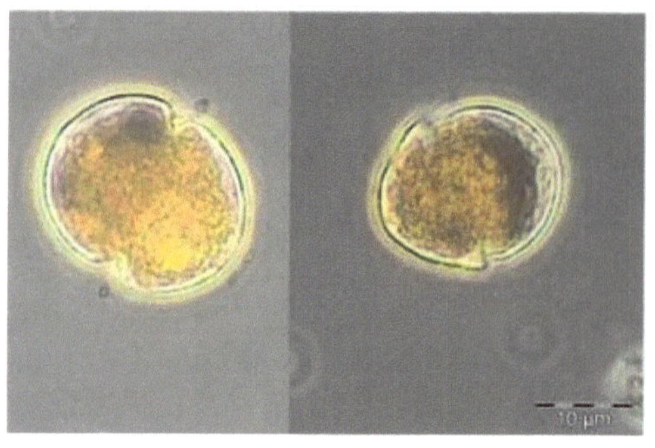

Figure 3-29. *Karlodinium veneficum* from station OMBMPJ1, 20 m depth, on 14 May 2016. Photo: S. Busch (IOW).

nucleus (Harry W. Rathbone, 2021) (Bruce A. Curtis, 2012). Scientists are debating whether the plastids in *D. baltica* and *K. veneficum* originated from a single symbiotic event or several. This debate is complicated because dinoflagellates often gain and lose plastids throughout their evolution, making it hard to trace the origin of these plastids. Both *K. veneficum* and *D. baltica* eventually lost their nucleomorph, and cryptophyte algae plastids replaced their original ones (Richard G. Dorrell, 2015).

Although uncommon, some dinoflagellates, such as *Gymnodinium aeruginosum* and *Gymnodinium acidotum*, are believed to practice kleptoplasty. Kleptoplasty is a process where a dinoflagellate consumes algae and breaks down the majority of the algae cells it engulfs, yet retains and preserves the algae's plastids. As a free-floating plastid, it continues to photosynthesize within the host, providing nutrients to the dinoflagellate. Unlike other endosymbiotic events previously mentioned, the retained plastid is not integrated biochemically or genetically into the host cell during kleptoplasty. The plastid also does not reproduce and will degrade within a few days to weeks. The host must continue consuming

more plastid-containing algae to restock their plastid stores (Tove M. Gabrielsen, 2011). Kleptoplasty may provide an evolutionary advantage to the host organism and create disadvantages. Kleptoplasty can provide stability in an environment where prey availability is inconsistent— allowing the host to survive long stretches of scarcity. However, this arrangement can lead to a dependency on specific prey types that contain suitable plastids.

In conclusion, the evolutionary narrative of dinoflagellates is complex and multifaceted, encompassing a remarkable journey through primary, secondary, and tertiary endosymbiotic events. These events have shaped their biology, allowing them to adapt and survive through drastic trait changes, including the loss and acquisition of plastids. The secondary endosymbiosis exhibited by *Kryptoperidinium foliaceum*, with its unique maintenance of endosymbiotic diatoms, represents an evolutionary marvel that illustrates the intricate relationships dinoflagellates can form. This symbiosis and the practice of kleptoplasty by species like *Gymnodinium* highlight the dinoflagellates' versatile strategies for photosynthesis and survival. Such adaptations grant them access to a broader ecological niche and enable them to thrive in diverse environmental conditions. Despite these advances, the reproduction mechanism by which dinoflagellates retain these symbiotic and kleptoplastidic relationships across generations remains a fascinating mystery, prompting further inquiry into their complex life cycles.

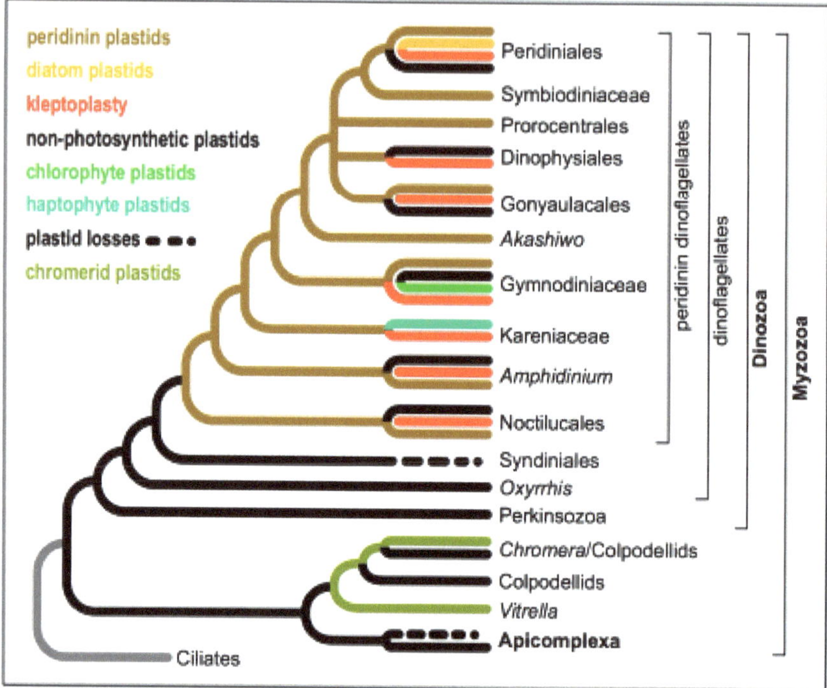

Figure 3-31. This image illustrates the complex plastid and pigment lineages within dinoflagellate phylogeny (Ross F Waller, 2017)

3.8 Ancient Glaucophyta

Glaucophyta is a distinct division within the *Archaeplastida* supergroup, which includes red and green algae. This division of algae provides a unique glimpse into the intermediary evolution from the initial primary endosymbiotic event to the fully evolved red and green algae. *Glaucophyta* plastids possess many genes shared mainly with other *Archaeplastida* members, yet they also retain a subset of genes unique to their lineage. These genes include those related to essential cellular processes like the assembly of photosystem II and the synthesis of quinolinate (Francisco Figueroa-Martinez, 2018). The conservation of these functions allows researchers to observe how other members of the Archaeplastida groups, such as red and green algae, may have transformed their genome from that of free-living cyanobacteria to the genetic profile of modern plastids (Amna Komal Khan, 2020).

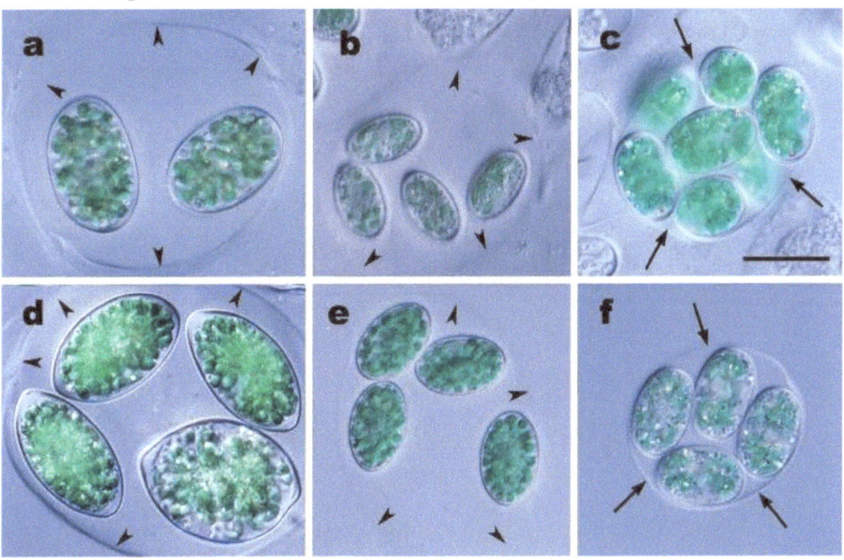

Figure 3-32
(a) *Gonium geitleri*
(b) *Gonium nostochinearum*
(c) *Gonium incrassata (Lemmerm.)*
(d) *Gonium oocystiformis*
(e) *Gonium miyajii*
(f) *Gonium bhattacharyae*
These are colonies of *Glaucophyta* species. The arrows highlight the enclosed mother cell wall that encompasses each colony (Toshiyuki Takahashi, 2016).

The glaucophyte plastids, known as cyanelles, are particularly compelling because they closely resemble their cyanobacterial ancestors, complete with a peptidoglycan layer between their two surrounding membranes, a feature absent in red and green algae plastids. This ancestral characteristic indicates that *Glaucophyta* may represent a transitional stage in plastid evolution, encapsulating the change from a free-living cyanobacterium to an integrated cellular organelle. Furthermore, the presence of a nucleomorph in some closely related algae suggests that glaucophyte plastids could reveal insights into the early stages of genome reduction associated with endosymbiosis, providing a living snapshot of the processes that may have occurred in the ancestors of red and green algae as their plastids became more specialized (Olivier De Clerck, 2012).

Moreover, the study of *Glaucophyta* contributes significantly to our understanding of the evolutionary trajectory and the diversification of autotrophic eukaryotes. As we unravel the genomic intricacies of *Glaucophyta*, we find clues about the early evolutionary pressures that shaped the integration of photosynthetic organelles into eukaryotic life. This knowledge not only sheds light on the origins of photosynthesis in eukaryotes but also helps to piece together the complex history of how plastids, which fundamentally altered the direction of life on Earth, evolved into the diverse forms we see today within the Archaeplastida supergroup (M. Vesteg, 2009).

3.9 Independent Photosynthetic Acquisition

The discussion of algal evolution would not be complete without mentioning the genus *Paulinella*, which provides a unique perspective on the evolution of photosynthetic organelles. Unlike the plastids that define the Archaeplastida lineage (red, green, and *Glaucophyta*), *Paulinella*

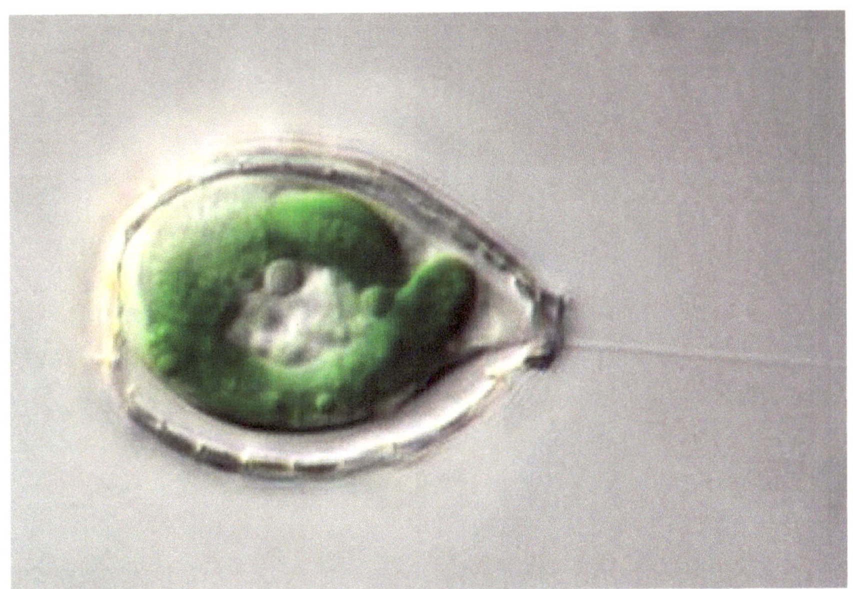

Figure 3-33. *Paulinella chromatophora*. Image: Eva Nowack, Univ. zu Köln.

chromatophora harbors chromatophores derived from a relatively recent and independent primary endosymbiotic event. This secondary assimilation of cyanobacteria occurred merely 90 to 140 million years ago, contrasting sharply with the more ancient event over 1.6 billion years prior that gave rise to the plastids found in algae and plants(Gabr et al., 2020). It is important to note that their unique plastid origin means they are not considered algae. The chromatophores of *Paulinella* species initially believed to be mere endosymbiotic cyanobacteria, have been revealed through genomic analysis to encode approximately 1,000 genes, highlighting a significant degree of autonomy. Such genetic endowment implies a complex symbiotic relationship where the chromatophore contributes extensively to the photosynthetic machinery of the host amoebae(Sato et al., 2020).

The study of *Paulinella* underscores the possibility of multiple origins of cellular organelles through endosymbiotic events, thereby enriching our understanding of eukaryotic evolution. As such, Paulinella is a testament to the evolutionary innovation possible within the eukaryotic domain, where endosymbiotic events continue to drive the emergence of novel cellular complexities.

3.10 Chapter Summary

The vibrant narrative of algal evolution unfolds across the vast tapestry of our oceans, a testament to the remarkable adaptability and survival of marine life. Algal endosymbiosis sits at the heart of this evolutionary saga, showcasing nature's ingenious collaborations that have enabled algae and cyanobacteria to pioneer life in various marine and freshwater habitats. This chapter has charted the remarkable journey of these organisms, from their earliest endosymbiotic origins to the intricate relationships they uphold with non-photosynthetic organisms, constantly reinventing themselves in the face of the ocean's relentless threats. The pathways of endosymbiosis, from the assimilation of red and green algae plastids to the enigmatic process of kleptoplastidy, highlight the evolutionary ingenuity these algae demonstrate.

The ancient origins of algal endosymbiosis sowed the seeds for the diversity we observe in today's marine ecosystems. With their inclination for forming novel symbiotic bonds, present-day dinoflagellates illustrate the evolutionary drive to forge partnerships that transcend mere survival, enabling a collective to thrive in the ever-changing seascape. Events like plastid capture and exchange have contributed to the complexity of algal evolution and influenced the development of other marine organisms that rely on these photosynthetic pioneers.

As we delve into the marine world, we are continually awed by the evolutionary prowess of these microscopic giants. The resilience of algae, mirrored in endosymbiosis, serves as a paradigm for life's grand design—where innovation meets necessity, and adaptation is the currency of existence.

In the upcoming chapters, we shift our focus to examine how algae and cyanobacteria have evolved and shaped the evolutionary paths of a host of non-photosynthetic organisms. We will explore the biochemical and genetic underpinnings of these symbiotic interactions, uncovering the mechanisms by which algal endosymbionts influence the metabolic landscapes of their hosts. Prepare to delve into the molecular genesis of these relationships as we seek to comprehend how the exchange of genetic material and the orchestration of metabolic pathways have sculpted the evolutionary narrative of life beneath the waves as well as on land.

Figure 4-1. Lichen Image by Jürgen from Pixabay

4. ALGAL SYMBIOSIS ACROSS KINGDOMS

4.1 Introduction

In the previous chapter, we embarked on a journey through the evolutionary history of algae, unveiling their transformative role in the web of life. Yet, it's imperative to recognize algae's profound influence in fostering symbiotic relationships that transcend the traditional boundaries of life's domains and kingdoms. Algae, cyanobacteria, and their plastid descendants have infiltrated every conceivable ecosystem, engaging in partnerships across diverse phylogenetic branches. These symbiotic unions mark the tale of algae's evolution, interweaving it with the narrative of life itself.

4.2 Lichen

Lichen represents a symbiotic organism system, often recognizable on tree bark, forest floors, or even inhabiting rocky facades. The complexity of lichens was unraveled by Simon Schwendener in 1867, whose microscopic studies revealed a dual composition of fungal structures and photosynthetic partners, algae or cyanobacteria (Honegger, 2000). Although Schwendener's discovery was foundational, the term "Symbiosis" was initially introduced in a biological context by Albert Bernhard Frank, who studied crustose lichens, and it was later popularized by

Figure 4-2. Simon Schwendener | Public Domain

Heinrich Anton de Bary, thus initiating a new understanding of these composite organisms (David L. Hawksworth, 2020) (Biography, 2023).

Figure 4-3.
General structure of lichen thallus: a. foliose b. fruticose c. crustose lichen
Attribution:
A) By Norbert Nagel, Mörfelden-Walldorf, Germany
B) By Jason Hollinger - Mushroom Observer
C) By Roantrum – Flickr
Diagram: (Beatriz Díaz-Reinoso, 2021)

Lichens are a testament to symbiosis, a relationship which manifests itself in diverse forms. (Lucie Vančurová, 2021). The fungus commonly develops a thallus structure, housing the algae or cyanobacteria, which may penetrate the algal cell walls to enhance nutrient exchange. Lichens can reproduce sexually, through the fungal partner, and asexually, by dispersing isidia[93] or soredia[94], ensuring the propagation of this symbiotic duo (Ines A. Aschenbrenner, 2016).

---

[93] **Isidia**: Outgrowths on lichens that contain both the fungus and algal cells, aiding in asexual reproduction and dispersion.
[94] **Soredia**: Small clusters of algal cells surrounded by fungal filaments, produced by lichens for asexual reproduction and dispersal.

Lichens hold immense ecological significance. They colonize environments often too harsh for other life forms, contribute to carbon uptake, mineral cycling, and energy flow within ecosystems, and provide food sources for various organisms. Lichens cover approximately 8% of the Earth's land surface, often dominating the biomass in forests, drylands, and tundras, underscoring their role in terrestrial ecology (Abas, 2021). Moreover, lichens are increasingly vital in assessing environmental health as bioindicators [95] in habitats threatened by pollution and other environmental stressors (David L. Hawksworth, 2020).

Figure 4-4. Lichen covered rocks on Heard Island tundra. Photo: Kate Kiefer

This symbiotic relationship also allows algae to inhabit terrestrial ecosystems, a significant leap from their ancestral aquatic origins. Through photosynthesis, algae offer carbohydrates, while the fungal partner contributes by absorbing and retaining water and minerals from the environment. The synergy within lichens exemplifies a successful adaptation to terrestrial life and reflects the broader ecological roles these symbioses play in nutrient-poor environments where they often prevail (Lucie Vančurová, 2021) (Ines A. Aschenbrenner, 2016).

---

[95] **Bioindicators**: Organisms or biological responses that reveal the health of an environment by signaling the presence of pollutants and the quality of habitats.

4.3 Coral

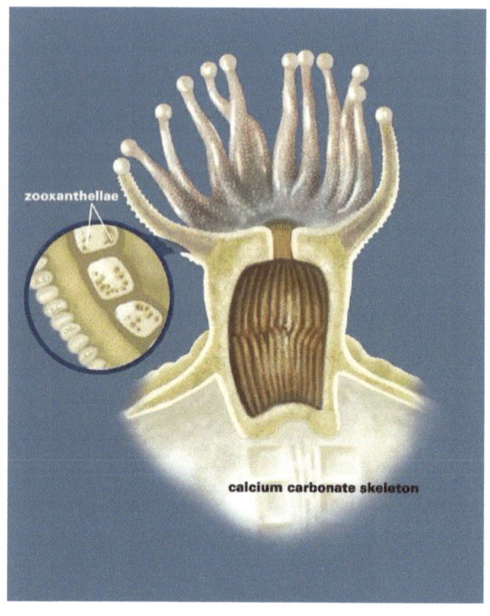

Figure 4-5. (Smithsonian Institution)

Coral reefs are integral to marine ecosystems, acting as the biological engines that produce the reef structure. The symbiotic relationship between corals and the dinoflagellate algae, known as *Zooxanthellae*, is fundamental to this process. The corals provide the algae with inorganic nutrients and a protected environment conducive to photosynthesis, while the *Zooxanthellae* reciprocate by supplying the corals with glucose, oxygen, glycerol, and amino acids synthesized during photosynthesis (Roth, 2014) (Andrew F. Torres, 2021). This mutualistic relationship is obligate, as neither organism can survive without the other. Stressful environmental conditions, such as ocean warming, can lead to the expulsion of the *Zooxanthellae* from the coral, resulting in coral bleaching and potential mortality of both organisms (Minjie Hu, 2023) (Nedeljka Rosic, 2015).

Figure 4-6. Tips of coral bleaching | U.S. Climate Resilience Toolkit

The *Zooxanthellae* reside within the coral's gastrodermis layer, shielding them from predators and environmental threats. Up to 90% of the products of photosynthesis are transferred to the coral, underlining the symbiosis's depth and the interdependence of these organisms. This endosymbiotic relationship is critical for the survival and proliferation of coral reefs. It plays a vital role in the larger context of marine biodiversity and the health of global oceanic ecosystems (Andrew F. Torres, 2021).

The coral-algal symbiosis has endured for over 200 million years. It is pivotal in forming and maintaining coral reefs, which are essential for an immense range of marine life. This partnership is also significant in climate change, as coral reefs are vulnerable to the impacts of ocean acidification[96] and rising temperatures, which can disrupt the delicate balance of this symbiotic relationship. Consequently, the resilience and adaptability of coral reefs in the face of climate change may hinge on the stability of the coral-algal symbiosis, underscoring its ecological importance (Fitzgeorge-Balfour, 2021) (Tamar L. Goulet, 2021).

4.4 Malaria

Plasmodium falciparum, the parasite which causes malaria, is a fascinating example of endosymbiotic evolution. This parasite harbors a unique plastid known as an apicoplast, a relict organelle originating from an endosymbiotic event with a eukaryotic alga (Chakraborty, 2016). Contrary to the expected evolutionary selection for energy-producing photosynthesis, the apicoplast is non-photosynthetic and functions primarily in synthesizing essential compounds such as amino acids, fatty

---

[96] **Ocean acidification**: The ongoing decrease in the pH of the Earth's oceans, caused by the uptake of carbon dioxide (CO2) from the atmosphere.

acids, and proteins crucial for the parasite's survival and virulence (Geoffrey Ian McFadden, 2017) (Morgan E. Milton, 2016).

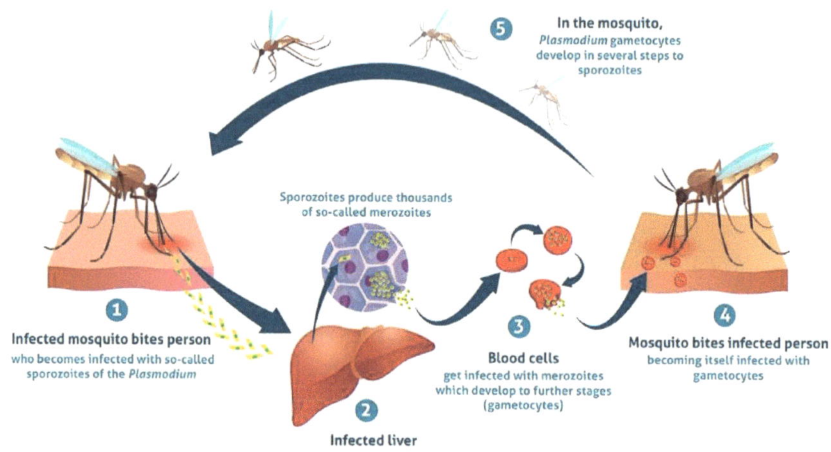

Figure 4-7. Malaria Transmission and Infection Cycle | Biogents.com

The apicoplast's evolutionary journey from a photosynthetic origin is a remarkable adaptation. Over time, as *Plasmodium* adapted to a parasitic lifestyle within host cells, where photosynthesis was not feasible or necessary, the apicoplast evolved by losing its photosynthetic function. However, it retained and repurposed other metabolic pathways. These pathways are essential for synthesizing isoprenoid precursors, fatty acids, and heme, which are vital for the parasite's life cycle but absent in mammalian hosts. The isoprenoid synthesis pathway, in particular, is crucial for the malaria parasite's survival inside human red blood cells.

The impact of algal evolution, particularly the endosymbiotic events that gave rise to organelles like the apicoplast, extends deeply into biological systems and human health. Malaria, attributable to *Plasmodium*, is responsible for nearly half a million deaths annually worldwide (Chakraborty, 2016). The unique features of the apicoplast, including its prokaryotic origin and absence in human cells, make it an excellent target for antimalarial drugs (Arwa Elaagip, 2022) (SooNee Tan, 2021). Developing therapies that disrupt the apicoplast's functions offers a means to combat malaria and reduce its significant burden on global health (Liting Lim, 2010) (Ellen Yeh, 2011).

4.5 Cyanobacteria and Nitrogen Fixation

Cyanobacteria are notable for their capacity to establish symbiotic relationships with various plants, including bryophytes—a group of non-vascular plants comprising mosses, liverworts, and

Figure 4-8. *Anthoceros sp.* Hornworts Photo by Bramadi Arya

hornworts. Hornworts, in particular, maintain a consistent symbiosis with cyanobacteria, especially from the genus *Nostoc*, housed within specialized cavities in the hornwort's tissues to fix atmospheric nitrogen (David G. Adams P. S., 2008). This relationship is mutually beneficial: the hornwort provides the cyanobacteria with a secure environment and carbohydrates from photosynthesis, while in return, the cyanobacteria supply the hornwort with bioavailable nitrogen, essential for the plant's growth in nutrient-scarce conditions (Consolación Álvarez, 2023).

Figure 4-9. *Gunnera manicata*, photographed near the church at St Just in Roseland in Cornwall.
Photo by Tom Oates.The original uploader was Nabokov at English Wikipedia

Figure 4-10. *Azolla caroliniana*
By Kurt Stüber CC BY-SA 3.0

Cyanobacteria also enter into symbiotic relationships with other plant species, including non-vascular and vascular plants such as the water fern *Azolla* and the angiosperm *Gunnera* (Brenda S. Prattel, 2021) (Sang-Moo Lee, 2021). Algae, while not involved in nitrogen fixation, can affect the microhabitat by colonizing the surfaces of bryophytes and thus potentially influencing the environment to support the presence and activity of nitrogen-fixing cyanobacteria (W.D.P. Stewart, 1983).

These symbiotic associations are crucial for the nitrogen cycle in ecosystems where bryophytes are familiar, like moist forests and wetlands (David G. Adams P. S., 2012). The interdependencies between bryophytes, algae, and cyanobacteria represent ancient evolutionary relationships vital for their survival and ecological success, particularly in nutrient-poor environments where they are often a dominant component of the flora (Bahareh Nowruzi, 2021).

4.6 Chapter Summary

In weaving the narrative of symbiosis throughout this chapter, we have journeyed from the lichen-covered forest floors to the vibrant, life-sustaining coral reefs and delved into the microscopic world where the battle against malaria is fought. The evolutionary tapestry of algae and cyanobacteria, with their capacity for symbiotic relationships, has demonstrated a remarkable breadth of influence across various ecological and biological domains.

The lichens, those complex organisms adorning tree bark and rocky facades, exemplify the harmonious coexistence of fungi and photosynthetic partners, a symbiosis that revolutionized our understanding of ecological interdependence. This partnership enables algae to thrive on land and fortifies lichens to become ecological sentinels and significant contributors to terrestrial ecosystems.

In the marine realm, the symbiotic bond between coral and algae, particularly the dinoflagellate *Zooxanthellae*, forms the cornerstone of coral reef ecosystems. These alliances are a testament to the power of symbiosis in creating structures that are not only biological wonders but also critical habitats teeming with marine life. The mutualistic relationship is so integral that environmental stresses causing disruptions in this alliance can lead to coral bleaching, signifying the fragility of these ecosystems in the face of climate change.

Turning to the microscopic, the *Plasmodium* parasite's retention of a plastid from its algal ancestry exemplifies how endosymbiotic events have shaped ecological systems and the fabric of life, impacting human health globally. The apicoplast's role in the parasite's life cycle presents a unique avenue for therapeutic interventions, highlighting the broader implications of these symbiotic events.

Finally, we explored the cyanobacteria and their pivotal role in nitrogen fixation within ecosystems. Their ancient symbiotic relationships with bryophytes, ferns, and angiosperms are foundational to the nitrogen cycle, particularly in nutrient-poor environments. These partnerships enhance the ecological success of their host plants and, by extension, the ecosystems they inhabit.

Each of these symbiotic relationships - from lichens to coral reefs, from the battlegrounds of malaria to the nitrogen-fixing bryophytes - is a microcosm of coexistence reflecting the grandeur of life's interconnectedness. They are the threads that weave the rich tapestry of life, demonstrating resilience and adaptability across the planet's varied landscapes and seascapes.

In conclusion, the stories of symbiosis in this chapter are more than evolutionary curiosities; they are fundamental forces shaping the biosphere. As we continue to explore these complex interactions, we deepen our appreciation for the intricate and resilient web of life that algae and cyanobacteria have helped to weave. It is a web that sustains, challenges, and ultimately connects us to the myriad forms of life with which we share our planet.

5. MEMBRANE DYNAMICS AND EVOLUTION

5.1 Vesicular formation

Let us briefly discuss phagocytosis to understand the process that leads to the sustained preservation of ingested organisms into endosymbiotic relationships. Phagocytosis[97] is a method of nutrient ingestion by which an organism can engulf the target, typically another cell or nutrient, using its cell membrane. In this process, the cell's membrane extends inwards into the cell, enveloping the target (Eileen Uribe-Querol, 2020). During this process, the inward-moving membrane fully engulfs the target, and the membrane-enclosed vacuole surrounds the target and isolates it from the internal biology of the phagocytic cell. It is at this point that enzymes can begin the digestive process (Natalya Yutin, 2009).

5.1.1 Microbial Prey detection

The phagotrophic microbe must initially sense its presence to capture a prey microbe. Prior to making contact with a potential prey, the phagotrophic microbe may use either chemotaxis[98], mechanoreception[99], or a combination of both. Chemotaxis is detecting gradients in the

---

[97] **Phagocytosis**: Cellular process of engulfing particles to form an internal compartment called a phagosome.

[98] **Chemotaxis**: Movement of a cell towards chemical stimuli, often seen in cells approaching nutrients or escaping toxins.

[99] **Mechanoreception**: Sensory transformation of mechanical forces into biochemical signals, enabling cells to detect changes in their environment.

concentration of specific metabolites[100] and waste that microbes expel (Ilan Chet, 1971). This process can be surprisingly complex, allowing microbes to detect the direction of a metabolite's source by interpreting the gradient across its receptor membrane (Johannes M. Keegstra, 2022). Figure 5-1 shows the biochemistry behind chemotaxis in *E. coli*. Bacteria. Chemical receptors on the surface of the bacteria membrane will detect a nutrient or harmful compound. The detection of these compounds will cause a cascade of biochemical reactions, which will result in the release of compounds that affect the rotation direction of the flagella in the bacteria. Depending on the type and intensity of the compound detected, the rotation will cause the bacteria to either move towards or away from

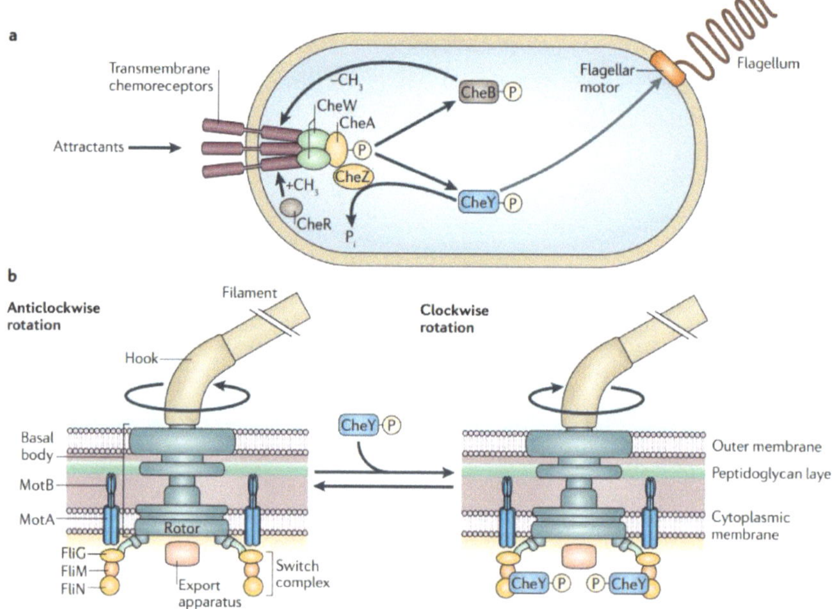

Figure 5-1. Escherichia coli Bactria (Steven L. Porter, 2011)

[100] **Metabolites**: Small molecules involved in the metabolic processes within organisms, often signaling molecules or energy sources.

the source of the compound. It is important to note that this is the function in *E coli*, And the exact mechanisms for reception and motility will vary based on the capabilities of the cell. However, it is essential to note that this is a purely biochemical process with no actual decision-making on the part of the cell (Steven L. Porter, 2011).

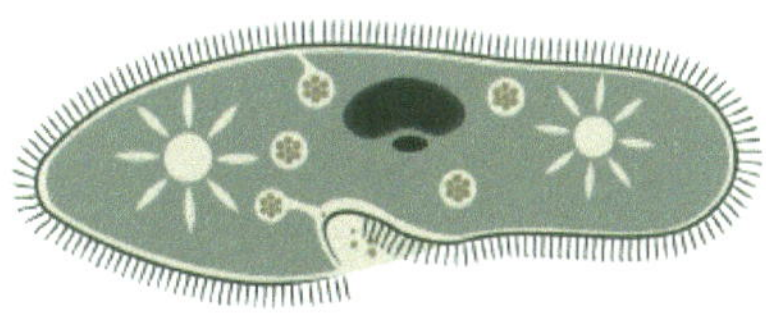

Figure 5-2. Paramecium are covered in cilia. The cilia help them move in one direction in water. | Study.com

Mechanoreception, as the name suggests, is a more mechanical approach to prey detection. Mechanoreception involves using specialized appendages, including cilia [101], to detect changes in the direction and intensity of currents to detect the presence of prey nearby (Carmen M. Cromer, 2021). The phagotrophic cell will close in on and contact the target prey. At this point, the microbe will "decide," in a biochemical sense, whether to consume the prey. This so-called "decision-making process" involves receptors [102] on the surface membrane that detect general or specific biochemical signatures. These receptors are specialized structures that respond to particular carbohydrates or proteins on the prey's surface membrane. This contact will then trigger the next step in this process, phagocytosis. It is essential to point out that although most common, these are not the only ways of prey detection observed. Some phagotrophic microbes may use specialized photoreceptors [103] that can use light intensity to detect prey, along with other microbes that use

---

[101] **Cilia**: Hair-like structures on cells that move fluid or detect environmental stimuli, aiding in locomotion and sensing.

[102] **Receptors**: Protein molecules on cell surfaces or within cells that bind to specific substances, initiating cellular responses.

[103] **Photoreceptors**: Cells or structures in organisms that detect and respond to light, influencing behaviors like prey detection.

temperature gradients or electric fields for prey detection (Netra Pal Meena, 2017) (Jean-Baptiste Raina, 2019).

5.1.2 Phagocytosis

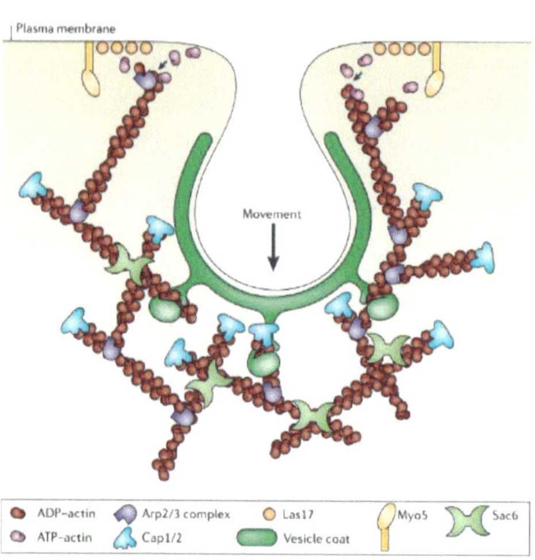

Figure 5-3. This graphic shows how the actin cytoskeleton filaments shapes the cell membrane to form the vesicle | (Marko Kaksonen, 2006)

Once contact is made with a recognized prey, the phagotrophic microbe will use specialized molecules called ligands [104] to facilitate the adhesion of the prey's membrane to the predator's cell membrane (Christa Decker, 2021) (Qingxian Lu, 2010) (Yan Yang, 2022). During this phase, the molecular switch ATPase [105] activates, specifically the Rac and Cdc12 ATPases. These molecules, when triggered, rearrange the actin cytoskeleton[106] (Nicolas Tapon, 1997) (Szczepanowska, 2009). The cytoskeleton is a network of actin protein filaments used to manipulate the shape of the cellular membrane, among other functions.

[104] **Ligands**: Molecules that bind to receptors, triggering a response from the cell, critical for cell signaling.

[105] **ATPase**: Enzymes that hydrolyze ATP, releasing energy for cellular processes, including muscle contraction and nerve impulse propagation.

[106] **Actin Cytoskeleton**: Network of actin protein filaments providing structural support and enabling cell movement and shape change.

The activation of ATPase initiates a series of biomolecular signals, which in turn stimulate the rapid polymerization [107] of actin monomers[108] into filamentous structures, forming the actin cytoskeleton (Pekka Lappalainen, 2022). In turn, this forces the cell membrane to protrude outward. In a phenomenon called treadmilling[109], actin subunits are strategically added and removed from the tip of the pseudopod [110] or protruding membrane, assisting the movement of the protruding membrane (Marie-France Carlier, 2017). This treadmilling coordinates the enveloping of the prey organism within the extending membrane. The process results from a complex and continuous feedback system, with dynamics and other principles working together to pinch off the phagosome [111] within the predatory cell.

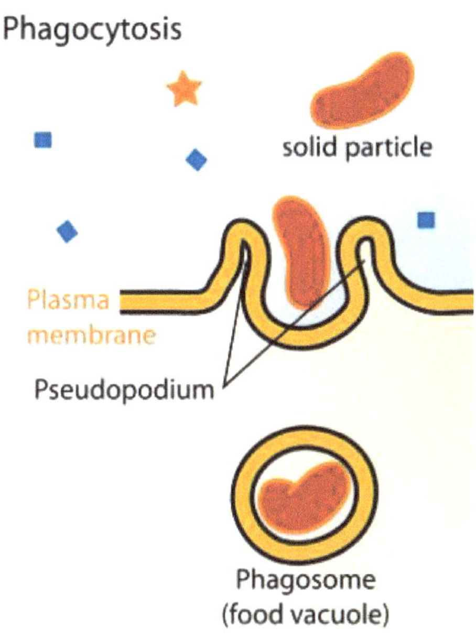

Figure 5-4. Adapted from image By Mariana Ruiz Villarreal LadyofHats

---

[107] **Polymerization**: Process of reacting monomer molecules together to form polymer chains, essential in the formation of cellular structures.

[108] **Actin Monomers**: Individual actin molecules that polymerize to form long chain filaments, integral to cell motility and structure.

[109] **Treadmilling**: Dynamic process where actin filaments grow and shrink, allowing cellular structures like pseudopods to extend and retract.

[110] **Pseudopod**: Temporary arm-like projection of a cell used for movement or engulfing particles during phagocytosis.

[111] **Phagosome**: Intracellular vesicle formed by the engulfment of a particle by a cell, leading to digestion of the particle.

Phosphoinositides[112], such as PI(3,4,5) and P3, will become enriched[113] at the location of phagosome creation (Valentin Jaumouillé, 2020). This process aids in the fusion adhesion[114] of the phagosome membrane. The final pinching off of the phagosome is triggered when GTPase[115] molecules detect the contact made by the tips of the pseudopods as they engulf the object. SNARE proteins[116] trigger the final reformation of the membrane to ensure proper phagosome closure (Kirsty M. Hooper, 2022).

5.1.3 Phagosome Maturation

Post-formation maturation[117] must occur to initiate digestion and nutrient intake into the predatory cell. This newly formed phagosome is like a soap bubble floating within the cell. The intravacuolar[118] media is of similar chemical composition to the extracellular[119] space. The inside

---

[112] **Phosphoinositides**: Lipids in cell membranes involved in signaling pathways, especially in the formation of phagosomes.

[113] **Enriched**: In biological terms, having a higher concentration of a particular molecule, often relating to a localized cellular area.

[114] **Fusion Adhesion**: Process where two membranes stick and merge, critical for forming cellular structures like phagosomes.

[115] **GTPase**: Enzyme that hydrolyzes GTP, involved in signal transduction and regulation of cell growth.

[116] **SNARE Proteins**: Family of proteins that mediate the fusion of cellular membranes, crucial for vesicle trafficking.

[117] **Maturation**: Biological process where cells or organelles develop to become fully functional, as in phagosome maturation.

[118] **Intravacuolar**: Inside a vacuole or phagosome within a cell, where intracellular digestion takes place.

[119] **Extracellular**: Located outside a cell, in the surrounding body fluids or tissues.

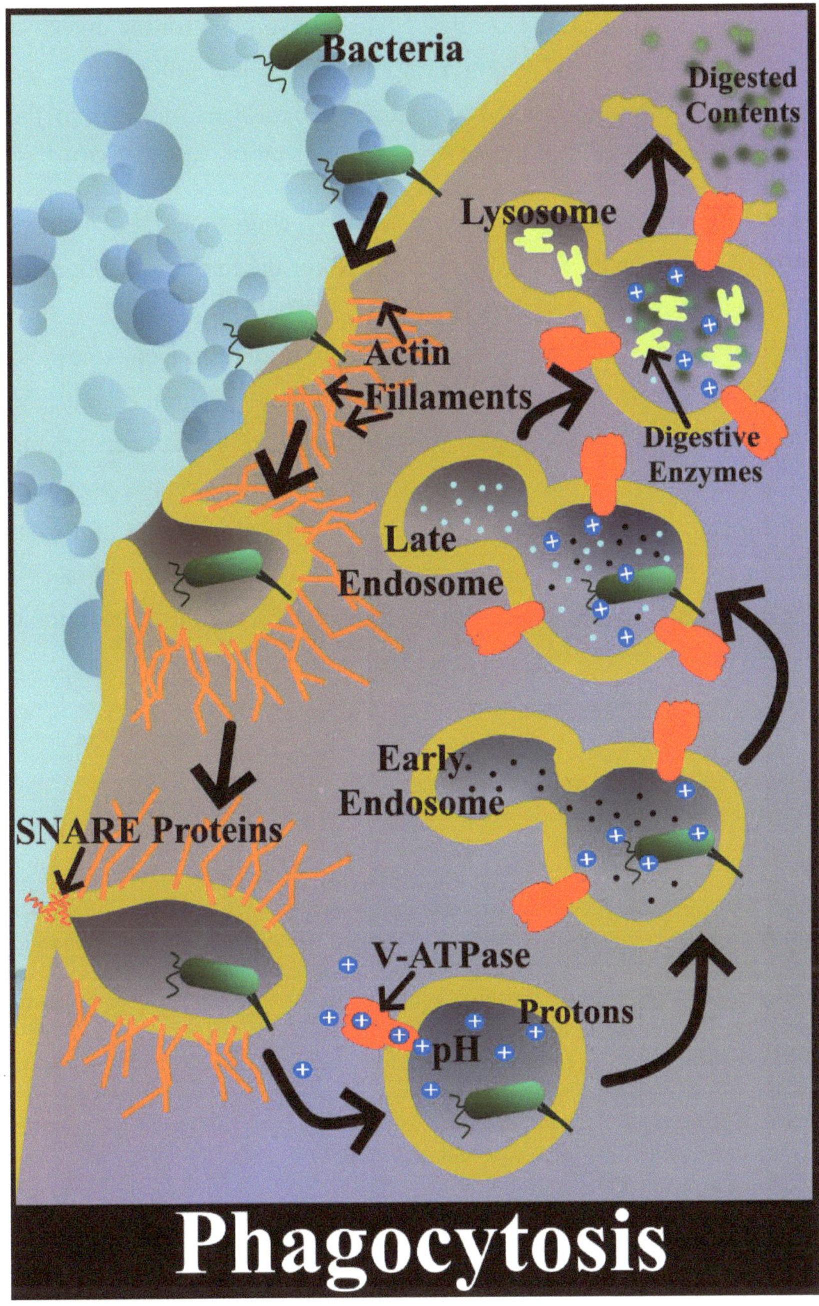

Figure 5-5. Phagocytosis By Victor M. Rodriguez 2023

lacks any digestive enzymes[120] and is unsuitable for the digestive processes. V-ATPase[121] & proton/proton pumps[122] are embedded in the phagosome membrane, transporting positive hydrogen ions[123] into the vacuole and rapidly acidifying the interior (Michael P Collins, 2020).

The phagosome first fuses with early endosomes[124] to receive membrane components and proteins required for the further development of the phagosome (Anne-Marie Pauwels, 2017). The phagosome then fuses with late endosomes[125], which contribute to digestive enzymes and cause further lowering of pH levels[126] (Marina Santic, 2008). Finally, the phagosome fuses with lysosomes[127], organelles filled with the enzymatic cocktail needed to process all the components of the captured prey into useful molecules (Jenny A. Nguyen, 2021). This final stage of fusion results in the creation of a phagolysosome. The phagolysosome[128] can be considered analogous to a temporary stomach of the cell, an extremely acidic environment with the enzymes required to reduce lipids[129], proteins,

---

[120] **Enzymes**: Proteins that catalyze biochemical reactions in the body, essential for metabolism and other cellular processes.

[121] **V-ATPase**: Enzyme that transports protons across membranes, acidifying intracellular compartments like phagosomes.

[122] **Proton/Proton Pumps**: Proteins that move protons across a membrane, establishing a gradient used in energy production and cellular homeostasis.

[123] **Positive Hydrogen Ions**: Protons, which contribute to acidity in a solution, important in cellular compartments like lysosomes.

[124] **Early Endosomes**: Membrane-bound compartments that sort endocytosed material, critical for phagosome maturation.

[125] **Late Endosomes**: Organelles that receive digested material from early endosomes and prepare it for lysosomal degradation.

[126] **pH Levels**: Measure of acidity or alkalinity in a solution, with lower pH being more acidic, important in lysosome function.

[127] **Lysosomes**: Cell organelles containing digestive enzymes, involved in breaking down waste materials and cellular debris.

[128] **Phagolysosome**: Fusion of a phagosome with a lysosome, resulting in an enzymatic environment for digestion of engulfed material.

[129] **Lipids**: Fats and fat-like substances that are important components of cell membranes and sources of energy.

polysaccharides[130], and nucleic acids[131] to their essential constituent components (Jenny A. Nguyen, 2021).

Once the degradation process is complete, the resulting compounds are either utilized in further metabolic processes or expelled back into the extracellular media if not useful. The organism disassembles the phagolysosome using vesicular trafficking[132], and all the enzymes and membrane components are recycled and reused (Roni Levin-Konigsberg, 2021).

5.2 Processes of Digestion Prevention

Several methods are proposed by which the engulfed symbiont cell survives or prevents digestion altogether. Some suggested methods involve biochemical signaling and modification, while others involve simple physical defense aids via barriers. This initial prevention of cellular degradation is critical to ensure the microbe's survival to initiate the process of endosymbiosis & integration within the host cell.

---

[130] **Polysaccharides**: Carbohydrates formed by the linkage of multiple sugar molecules, serving as energy sources or structural components.

[131] **Nucleic Acids**: Biomolecules, including DNA and RNA, that store genetic information and are involved in protein synthesis.

[132] **Vesicular Trafficking**: Process of transporting materials within a cell via membrane-bound vesicles, integral to cellular logistics.

Figure 5-6. *Zooxanthellae* (brownish-green cells) inside translucent coral polyps. | osf.co.uk.

Cells have many ways they can recognize each other. This process usually involves detecting protein markers on the surface of their membrane or chemicals released by the Organism. In the context of endosymbiosis, there may be some instances in which the host cell may recognize the engulfed microbe and suppress its digestive processes. The microbe may seem harmless or as being "its self" and be actively protected. In the endosymbiotic relationship between Cyanobacteria *Synechococcus* and certain diatoms, digestion prevention involves complex membrane interactions that suppress lysosomal fusion. This process is crucial for the cyanobacterial endosymbiont to maintain its intracellular lifestyle within the diatom host. The specific loss or alteration of genes during the endosymbiont's genome evolution suggests a refined adaptation to the intracellular environment, preventing the usual endocytic pathway where lysosomes fuse with phagosomes to degrade their content. This strategic inhibition of lysosomal fusion facilitates the persistence and functioning of the endosymbiont, which is essential for nitrogen fixation within the host diatom (Christoph Kneip, 2008).

In some cases, digestion prevention through biochemical recognition might require a more active recognition mode. The dinoflagellate *Symbiodinium*, a critical endosymbiont in coral-dinoflagellate symbiosis, has evolved functions through adaptive selection that may actively prevent its digestion by coral polyps. Genomic analyses reveal genes related to transmembrane ion transport and the synthesis and modification of amino acids and glycoproteins. These are vital for the symbiotic relationship and could be responsible for signaling mechanisms that deter digestion. These molecular signals, possibly glycoproteins, might interact with the coral's physiology to discourage the digestive process, thus preserving the mutualistic relationship essential for coral reef productivity (Huanle Liu, 2018).

The process of biochemical recognition may result from an evolutionary memory[133] of some past interaction. Organisms with a shared evolutionary history may remember each other biochemically and treat each other more favorably. This evolutionary memory may explain how many dinoflagellates seem to have the propensity for acquiring new endosymbiotic relationships, even after losing plastids in the past. The history of serial endosymbiosis in dinoflagellates suggests that the host's biochemistry shifts to be more favorable to future events with every symbiotic interaction (András Szilágyi, 2020).

5.2.2 Nutritional and Environmental Triggers

There are instances when an organism may refrain from triggering digestive processes during nutrient deficiency to exploit the benefits of the consumed microbe. Experimental observations of ciliates

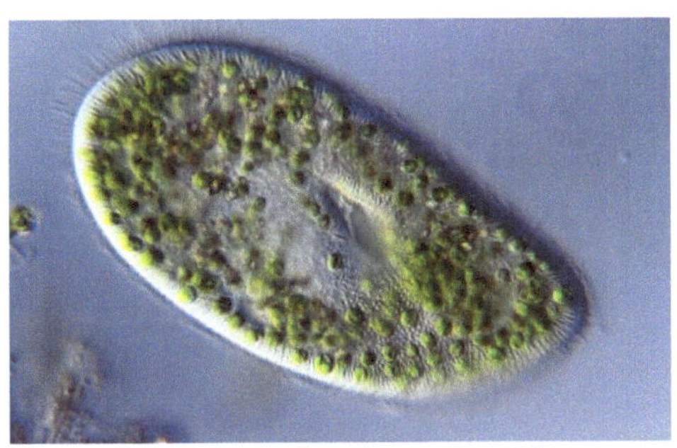

Figure 5-7. *Paramecium bursaria* with green zoochlorellae living inside endosymbiotically |y Anatoly Mikhaltsov

---

[133] **Evolutionary memory**: Genetically encoded behaviors or traits in an organism that result from adaptation to past environmental pressures.

consuming the green alga *Chlorella* have shown that some nutrient-deprived ciliates will choose to protect the *Chlorella* it has engulfed within to benefit from photosynthesis's byproduct. This temporary mutualism, a form of mixotrophy, allows metabolic interaction with foreign cells or organelles, partially enabling the ciliates to bypass typical heterotrophic digestion pathways. Unique adaptations in certain algal endosymbionts facilitate the establishment of this symbiosis and nutritional interaction while reducing their fitness for survival as free-living cells (Johnson, 2011).

This temporary mutualism is not limited to photosynthetic relationships. Some nitrogen-limited aquatic environments cause algal cells to ingest and then protect nitrogen-fixing bacteria within themselves to benefit from nitrogen fixation (Yuqian Tang, 2023).

5.2.3 Phagosomal Maturation Interference

As discussed previously, digestion requires the transition of the phagosome into a phagolysosome. Through Phagosome modification, however, the engulfed cell can interfere with and manipulate the composition of the phagosome to prevent fusion with heterotrophic endosomes[134], thus ending the digestive process (Anne-Marie Pauwels, 2017) (Jenny A. Nguyen, 2021).

Other maturation interference modalities involve the endosymbiont releasing effector proteins[135] into the host cell's cytoplasm

---

[134] **Heterotrophic endosomes:** Intracellular vesicles in heterotrophic organisms that digest and process internalized nutrients or symbionts.
[135] **Effector proteins:** Molecules secreted by cells that alter the behavior or properties of other cells, often used in immune responses and cellular communication.

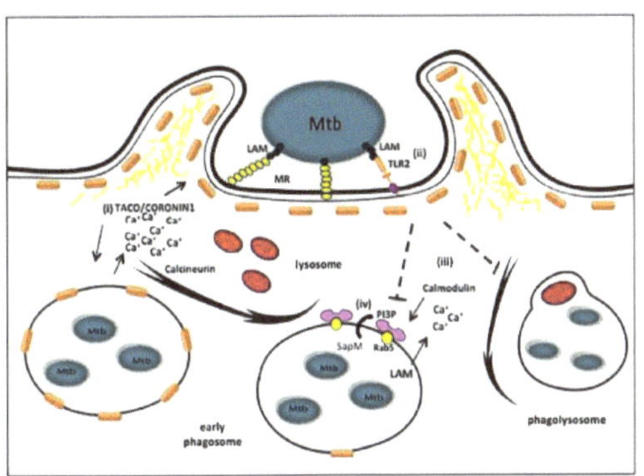

Figure 5-8. Key strategies of *M. tuberculosis* to evade host digestion.

to block or interfere with the biochemical signaling process of successful phagosomal maturation. Endosymbionts could also produce specific molecules that prevent the pH decrease in the phagosome or modify calcium concentrations to interfere with calcium signaling pathways (Jason M. Kinchen, 2008). Figure 5-9 describes the defense strategies *M. tuberculosis* employs to survive within host cells. The protein TACO disrupts the usual phagosome-lysosome fusion, a critical step for the bacterium's degradation. When the mannose receptor on a host cell binds with the bacterial wall component LAM[136], it causes TACO to remain on the phagosome membrane, triggering an influx of calcium ions. This influx is crucial because it activates calcineurin, inhibiting phagosome maturation. TLR2 receptor's[137] recognition of the bacteria also promotes TACO retention, aiding in this evasion mechanism. Paradoxically, LAM can also decrease calcium levels within the cell, activating calmodulin[138] that blocks PI3P[139], halting essential intracellular trafficking required for phagosome maturation. Additionally, the

---

[136] **Lipoarabinomannan (LAM):** A glycolipid found on the cell wall of M. tuberculosis, involved in immune evasion.

[137] **TLR2 Receptor:** A protein that recognizes pathogens and activates immune responses against them.

[138] **Calmodulin**: A calcium-binding messenger protein that transduces calcium signals by modifying various target proteins within the cell.

[139] **Phosphatidylinositol 3-phosphate (PI3P)**: A phospholipid that plays a role in membrane trafficking and signal transduction.

bacterium secretes SapM[140], a lipid phosphatase, which eliminates PI3P from phagosomes, preventing the docking of proteins necessary for phagolysosome formation, ultimately enabling *M. tuberculosis* to resist the host's attempts at digestion.

5.2.4 Physical Means of Prevention

In endosymbiotic relationships, specific physical structures of algae, like cell walls and the silica-based frustules of diatoms, may serve as undigestible barriers against degradation. Diatoms have unique siliceous cell walls—frustules—that are intricately designed and consistently replicated, providing physical protection and potentially deterring predators through chemical defense mechanisms. These frustules' complex shapes and symmetries contribute to the diatoms' resilience against digestion. Moreover, the biosilica acts as an effective pH buffer, possibly creating an unsuitable environment for the digestive enzymes of predators, contributing to the diatoms' survival and their ability to establish endosymbiotic relationships (Gross, 2012) (A. Reid, 2021).

5.2.5 Evolutionary Compromise

Whichever modality is employed to protect the symbiont from digestion, the presence of a new permanent resident poses A dilemma for the host. The host must decide if the means necessary to expel the intruder

---

[140] **SapM**: An enzyme released by M. tuberculosis that degrades PI3P, impeding phagolysosome formation.

is worth employing. In this context, autotrophic symbionts hold the endosymbiotic high ground. Autotrophs can better communicate their usefulness to the host through nutrient production, considering they produce nutrients and energy with relatively fewer inputs than heterotrophs.

5.3 Membrane Modification

5.3.1 Membrane Stability

Before any meaningful evolution occurs within the endosymbiont, the double membrane structure must be modified to ensure long-term stability. This process may include modifying embedded molecular structures or changing the permeability[141] of the membrane structure itself. The double membrane structure must act as a line of communication and transport between the endosymbiont and its host. The molecular markers embedded within the membrane are the first line of communication between the two organisms. The process of membrane modification requires communicating to the host that the undigested intruder is a beneficial member of this ongoing biomolecular machine. This process may involve the symbiont rapidly losing threatening protein surface markers while preserving and developing markers that signal mutual benefit to the host. This process may often involve biomolecular

---

[141] **Permeability**: Measure of a membrane's ability to allow substances to pass through it; crucial for nutrient exchange in symbiosomes.

"mimicry," wherein the symbiont organism develops antigen protein markers[142] that mimic those present in its host (Jason E. Cournoyer, 2022).

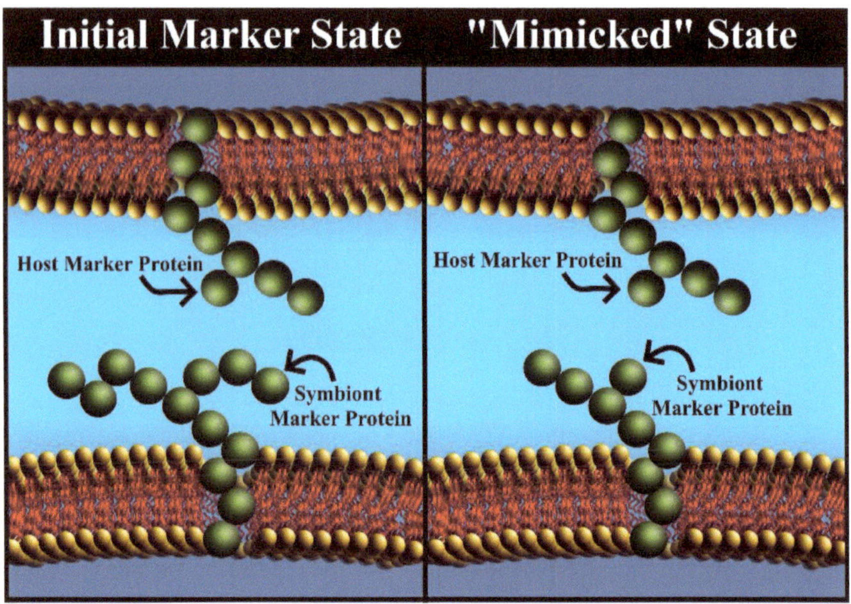

Figure 5-9. Diagram of protein marker mimicking by: Victor M. Rodriguez 2023

Contemporary examples of this protein modification may be part of several endosymbiotic relationships in which the symbiont retains its autonomy as an organism (Jeremy N. Timmis, 2004). Select diatom-Cyanobacteria relationships have been observed where the cyanobacteria are known to have modified antigen markers, allowing digestion avoidance and long-term mutualism (Artem A. Trofimov, 2019). This process is also evident in the Chlorella-like algae endosymbiotic relationship with *Paramecium bursaria* (Wernegreen, 2012). In the case of symbiotic partnerships between corals and dinoflagellates, this relationship and surface protein modification can be a less stable affair. During periods of environmental stress, it can cause a biomolecular

---

[142] **Antigen Protein Markers**: Proteins on cell surfaces that identify them to the immune system; may be altered in symbiosomes to avoid host detection.

Figure 5-10. The diatom *Epithemia turgida* with endosymbiotic cyanobacteria | Image Credit: Rex Lowe

mismatch in signaling, resulting in the termination of the endosymbiont relationship. Events like these can result in coral bleaching (Giada Tortorelli, 2022). These instances may serve as a window into the initial stages that eventually resulted in the development of plastids (Rebecca S. Meaney, 2020).

<u>5.3.2 Role of Membrane-Associated Enzymes</u>

Membrane-bound enzymes[143] are responsible for many processes involving cell membrane structure. In membrane modification for endosymbiotic establishment, these enzymes play a crucial role. One key role membrane-bound enzymes may play is in modifying the host's immune receptors, involving the degrading of phagosome receptors to evade detection by the host. Enzymes such as phospholipase D (PLD)[144] are responsible for the further expansion and remodeling of the membrane during the transformation of the phagosome into a symbiosome [145] (Matthias Corrotte, 2006). Phospholipase A2 (PLA2)[146] is responsible for

---

[143] **Membrane-bound Enzymes:** Enzymes attached to cell membranes, playing roles in transforming the phagosome to a symbiosome.
[144] **Phospholipase D (PLD):** Enzyme that modifies phospholipids, potentially aiding in the formation and maintenance of symbiosomes.
[145] **Symbiosome**: Specialized vesicle formed around symbiotic organisms, derived from the phagosome in endosymbiotic relationships.
[146] **Phospholipase A2 (PLA2):** Enzyme that hydrolyzes phospholipids at the sn-2 position, may be involved in symbiosome membrane remodeling.

hydrolyzation [147] of the sn-2 region [148] of phospholipids [149] fatty acids, which leads to the production of lysophospholipids [150], resulting in increased membrane permeability (Shibbir Ahmed Khan, 2023) (Deepti Dabral, 2017). Figure 5-12 illustrates the way the phospholipids in the viscous membrane become kinked when hydrolyzed by Phospholipase A2 as shown in the fluid membrane. This causes the phospholipids to spread apart, allowing for more permeability. This increase in permeability will become critical for nutrient exchange and reinforcement of the mutually beneficial arrangement.

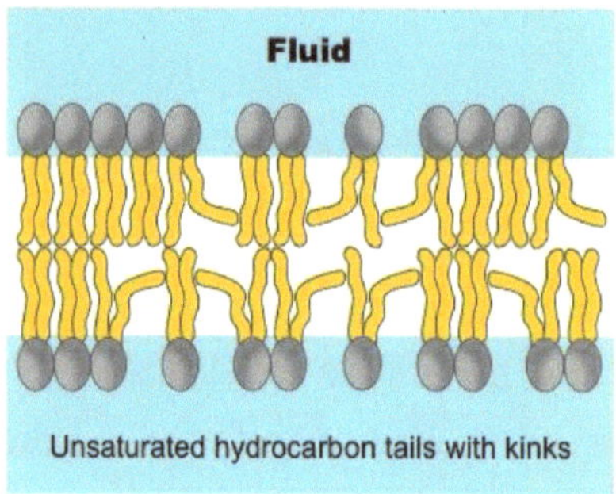
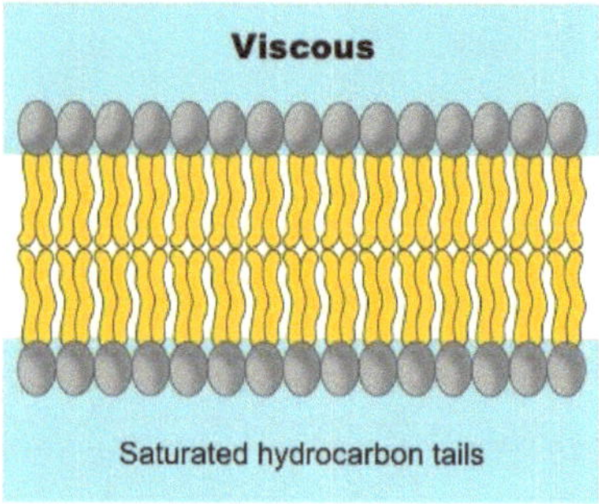

Figure 5-11. Credit: Bioninja.com.au

Being confined within a membrane provides the Symbiont with new challenges that need to be mediated to survive. Metabolic processes undertaken by the Symbiont result in the creation of Reactive Oxygen

---

[147] **Hydrolyzation**: Chemical breakdown of a compound due to reaction with water; involved in remodeling membranes during symbiosome formation.

[148] **Sn-2 Region:** Position in phospholipids where fatty acids are attached and targeted by enzymes like PLA2 during lipid metabolism.

[149] **Phospholipids**: Major components of cell membranes, their metabolism is key in the transformation of phagosomes into symbiosomes.

[150] **Lysophospholipids**: Byproducts of phospholipid metabolism, involved in membrane alterations during symbiosome formation.

Species (ROS)[151] , which may result in oxidative damage[152] to both the Symbiont and the host. Enzymes such as Superoxide Dismutase (SOD)[153] aid in the detoxification of the symbiosome, protecting both the Symbiont and host (Andrew T. Canada, 1989).

5.3.3 Development of Feedback Mechanisms

Robust cross-membrane communications are essential for the long-term survival of the newly developing endosymbiotic relationship. This establishment requires developing feedback mechanisms that keep the symbiont and host informed about each party's metabolic and nutritional needs (Ming He, 2019) (Minjie Hu, 2023) (Tingting Xiang, 2020). This feedback mechanism involves the development of protein and other molecular signals that may serve as biomolecular indicators (Catherine Kistner, 2002). These signals can inform the partners as to nutrient availability. This cross-membrane communication may also be helpful for the regulation of endosymbiont growth. The host may develop inhibitors [154] to regulate endosymbiont replication to prevent it from overconsuming host nutrients and endangering the symbiotic relationship (Tingting Xiang, 2020).

---

[151] **Reactive Oxygen Species (ROS):** Highly reactive molecules formed by the partial reduction of oxygen, can cause oxidative damage; may be regulated in symbiosomes.

[152] **Oxidative Damage**: Cellular damage caused by ROS, which symbiosomes may have mechanisms to minimize to preserve the endosymbiont.

[153] **Superoxide Dismutase (SOD):** Enzyme that mitigates oxidative damage by converting superoxide radicals into less harmful molecules, potentially protective in symbiosomes.

[154] **Inhibitors**: Substances that decrease, prevent, or block the activity of enzymes or other molecules

5.4 Nutrient and Metabolite Exchange

For a successful endosymbiotic relationship to develop and flourish, the partnership should be mostly mutualistic (Patrick J. Keeling J. P., 2017). Otherwise, such a relationship may become parasitic. Establishing this relationship requires the development of a wide spread of metabolic transport mechanisms to satisfy the needs of both the symbiont and host(Rafiqi et al., 2022). Maintaining osmotic homeostasis [155] between the two parties involves the development of aquaporins [156](Jiang et al., 2021)(Alleva et al., 2012). These are water transport channels within the membrane that allow for water movement and some small nutrients between the two parties.

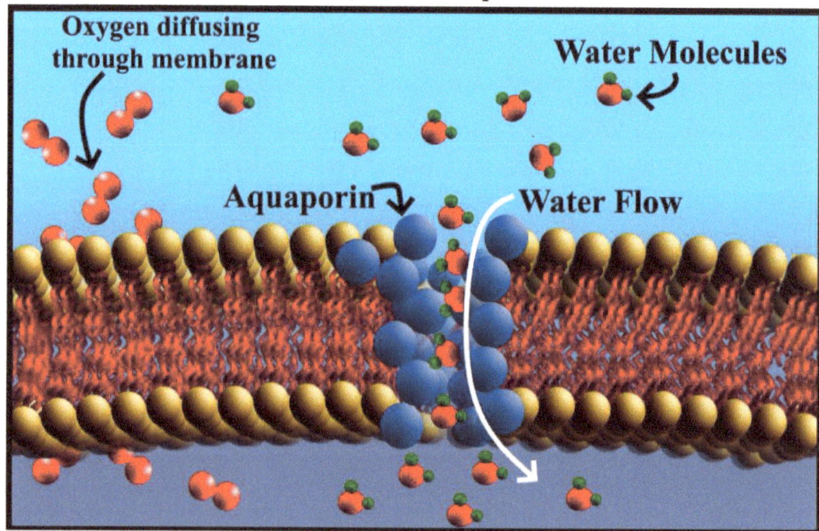

Figure 5-12. Illustration of aquaporins and oxygen diffusion through the symbiosome membrane | By Victor M. Rodriguez 2023

[155] **Osmotic homeostasis**: The maintenance of constant osmotic pressure in the fluids of an organism by controlling water and salt concentrations.
[156] **Aquaporins**: Proteins that form channels in cell membranes to facilitate the transport of water molecules in and out of cells.

Crucial for algal relationships specifically, Glucose transporters allow for the movement of glucose produced by the algal symbiont to the host for metabolizing(Xiang et al., 2020)(Burriesci et al., 2012)(Sproles et al., 2018). Amino acid transporters also provide a two-way facilitation of exchange between the parties(Dahan et al., 2015). These metabolite and nutrient transport mechanisms, among others, facilitate the mutualistic relationship between the two parties. This exchange can serve as a selective pressure mechanism for the further evolution of function specialization(Rafiqi et al., 2022). As an example, photosynthetic byproducts shared by the symbiont may eventually result in the reduction or loss of phototrophy by the host, as it may be less energy-intensive to maintain a favorable endosymbiotic environment for the photosynthetic symbiont instead(Flügge et al., 2016). Energy efficiency can be a powerful selective pressure for evolution.

This nutrient exchange develops an eventual symbiotic dependency in most cases. Both the symbiont and host will tend to lose functionality(Archibald, 2015)(Sibbald & Archibald, 2020). Specialization allows for both parties to streamline their respective metabolic and synthesis mechanisms. For example, plastids of cyanobacterial or algal origin have lost their ability to carry out most microbial functions except for photosynthesis. Meanwhile, the host organism must provide the plastid with inorganic materials for plastid maintenance and to facilitate photosynthesis.

Some critical nutrients exchanged between the host and symbionts are carbon compounds, nitrogenous compounds, and vitamins, among others(Cournoyer et al., 2022)(Iwai et al., 2019)(Dean et al., 2016). In photosynthetic endosymbiotic relationships, CO_2 transport from the extracellular media to the photosynthetic plastid is essential to continue photosynthetic functions(Cournoyer et al., 2022). The resulting O_2 from photosynthetic functions diffuses through the membrane into the host cytoplasm(Dean et al., 2016). From there, the oxygen is used in cellular respiration or expelled into the extracellular medium.

Metabolic and synthesis specialization coupled with nutrient exchange often further streamlines biochemical processes within the relationship. Some algal photosynthetic endosymbionts can uptake ammonia (NH_4^+) released by the host as a waste(Dean et al., 2016). The ammonia goes into producing glutamine, which is required to make some amino acids and glucose. This process, seen in coral/algal symbiosis, allows the coral to dispose of ammonia, a byproduct toxic to the coral in high concentrations.

The constant intermingling of metabolite and nutrient crosstalk further strengthens the endosymbiotic relationship between host and symbiont. The membrane modification process is vital in developing the pathways to facilitate this transfer. Membrane development can become an evolutionary struggle of life or death. The microbe engages in a dramatic race against time to modify the traditional digestive and immune system components to convince a host that symbiont preservation is preferable to elimination.

5.5 Membrane Evolution in Summary

As we explore the significance of endosymbiosis in algal evolution, the pivotal role of the evolving membrane becomes clear. This complex barrier, seemingly just a boundary, is a testament to nature's evolutionary genius. It evolved from a mechanism of death to a dynamic interface facilitating mutualistic exchanges.

Our journey through this chapter has emphasized the delicate balance needed in endosymbiotic relationships, whether parasitic or mutualistic. The vast array of metabolic transport systems that have evolved to cater to both the host and the symbiont's needs are critical to this balance. Aquaporins, glucose transporters, and amino acid transporters are just the tip of this intricate iceberg. These channels provide a conduit for nutrients and play a role in preserving osmotic homeostasis, a cornerstone for maintaining the internal environment of these partnerships.

While the initial evolution of the endosymbiotic relationship hinged on these transport mechanisms, the subsequent evolution involved an increasing specialization of functions. The algal symbiont and its host choreographed the dance of evolution by progressively streamlining their roles and cutting out redundancies. This journey saw plastids of cyanobacterial or algal origin narrowing down their roles predominantly to photosynthesis, even as the host evolved to nourish the symbiont with essential inorganic materials.

Yet, amidst this symphony of metabolic exchanges and functional specializations, the evolving membrane is an unsung hero. It adapted to facilitate the nuanced exchanges between the host and symbiont and underwent modifications to protect the symbiont from the host's digestive and immune systems. This protective role of the membrane underscores its significance, as it essentially convinces the host that preserving the symbiont is preferable to its digestion.

The chapter also delved deep into the biochemical intricacies underpinning these relationships. From the utilization of ammonia – a potentially toxic byproduct – by photosynthetic endosymbionts in amino acid production to the continuous tug and pull of nutrients. The CO_2 transport from the extracellular medium to the photosynthetic plastid was another instance, reflecting the depth of these interactions and exchanges.

In synthesizing our understanding, the journey through algal evolution reveals a remarkable narrative of collaboration and adaptation. At its core lies the membrane, a dynamic and evolving entity, shaping and being shaped by the endosymbiotic relationship. As we dive deeper into the genetic and biochemical aspects of this phenomenon, it's essential to carry forward the insights from this chapter: the story of life is as much about individual entities as it is about their relationships, and in this intricate dance of partnerships, evolution crafts masterpieces of collaboration and coexistence.

6. GENETIC AND MOLECULAR COMMUNICATION AND REGULATION

6.1 Genetic Exchange and Reduction

When one thinks of evolution, we often think of gradual change caused by random mutations. Single flips in genetic code cause specific traits to emerge and survive over time(Fitzgerald & Rosenberg, 2019)(Gregory, 2008). The truth is far messier than the grade school biology curriculum may lead you to believe. As mentioned throughout this book, evolution is often a convoluted web of divergent lineages converging again, forming endosymbiotic relationships that change both organisms' biology and evolutionary trajectory forever. In some cases, it gives rise to entirely new categories of life. Gradual expression and repression of genetic information are crucial to developing endosymbiotic relationships(Timmis et al., 2004). As functions and traits become obsolete or harmful to the endosymbiotic relationship, selective pressures will cause these traits to diminish while beneficial traits survive through evolutionary pressures(Fitzgerald & Rosenberg, 2019). However, this form of genetic evolution is a small part of a broader, more intricate exchange of information.

6.1.1 Endosymbiotic Gene Transfer (EGT)

EGT is a process by which the plastid membrane becomes permeable to genetic transfer. Although this process is not entirely understood, there are several ways that genetic transfer may occur. Small DNA-containing vesicles may bud off the original symbiosome, facilitating transport to the nucleus. Transportation may also be possible by developing transport protein complexes[157] that bind to the symbiont's DNA and allow for permeable transport across the symbiosomal membrane. An even more interesting hypothesis involves sacrificing a symbiont to facilitate gene transfer. The symbiont may have undergone either a planned or an accidental lysis event[158], which expelled the internal contents of the symbiosome into the cytoplasmic media. It should be noted that this mode of gene transfer would require the host cell to possess multiple instances of the same symbiosome to ensure proper restoration of the symbiotic relationship after the lysis event(Timmis et al., 2004).

Once the genetic material enters the host cytoplasm, it migrates to the host nucleus through specialized pore gateway complexes[159] or transport vesicles[160]. Once inside the host nucleus, the genes are integrated into the existing host DNA. The integration process will depend on the similarity of the foreign host gene to existing host genetic material(Timmis et al., 2004).

6.1.2 Homologous Recombination (HR)

When the new DNA strand is similar or "homologous" to the existing host strands of genes, the host can utilize homologous recombination as a means of genetic integration. HR is a repair mechanism

---

[157] **Transport protein complexes**: Protein assemblies that facilitate the movement of molecules across cell membranes.

[158] **Lysis event**: The process of breaking down a cell membrane, often to release genetic particles.

[159] **Pore gateway complexes**: Protein structures forming channels through membranes, allowing passage of molecules.

[160] **Transport vesicles**: Membrane-bound sacs that carry substances within and between cells.

essential in genetic maintenance(Ouyang et al., 2008). Without HR, an organism's DNA would eventually decay due to oxidative stress, radiation, or random degradation. The process of HR begins when a strand of host DNA undergoes a double-stranded cleavage or break. Enzymes process the DNA's damaged ends to create overhangs of single-stranded DNA. Nucleoprotein filaments[161] are formed at the strand ends when the Rad51 protein[162] binds to the overhangs. These nucleoprotein filaments seek out similar or complementary strands of DNA sequences(Bonilla et al., 2020). These strands do not have to be from the host cell genome; any DNA source can serve this purpose—even those of foreign organisms. When the Rad51 bound overhang filament finds a complementary strand (in the case of Symbiont Gene Exchange, the Symbiont DNA strand), the filament will 'invade' the strand. A process by which the single-strand filament will displace one strand of the complementary strand and bind with the other strand of foreign DNA.

Once the filament 'invades' the foreign DNA strand, DNA polymerases[163] recognize the paired-up host and symbiont DNA strands and begin the process of DNA synthesis[164] using the foreign strand as a template. The final result of this synthesis is a double helix strand that combines the host and symbiont strain(McVey et al., 2016).

---

[161] **Nucleoprotein filaments:** Structures formed by proteins and DNA, involved in DNA repair and replication.
[162] **Rad51 protein**: A DNA repair protein that helps in homologous recombination during gene transfer.
[163] **DNA polymerases**: Enzymes that synthesize DNA strands from nucleotides, crucial for DNA replication.
[164] **DNA synthesis:** The creation of new DNA molecules, fundamental for cell division and gene transfer.

<u>6.1.3 Non-Homologous End Joining (NHEJ)</u>

Just as in Homologous recombination, NHEJ starts with a double-strand breakage. However, with NHEJ, the foreign DNA strand in question does not have to be similar to the host DNA. ATM[165] and ATR proteins[166] detect the double-strand break (DSB) and activate the host's DNA repair pathways. Proteins Ku70 and Ku80[167] bind to the ends of the DSB, and DNA-PKcs[168] and ligase IV enzymes[169] initiate ligation[170](Zhao et al., 2020)(Pannunzio et al., 2018)(Menolfi & Zha, 2020). During this process, foreign strands of DNA, such as those belonging to a symbiont, can be mistakenly integrated into the host's genome, resulting in the hybridization[171] of the DNA. This process will likely cause additions, deletions, and mutations in the host genome, allowing for functions to be either gained or lost.

---

[165] **ATM Proteins**: Proteins that respond to DNA damage and initiate repair, key in maintaining genomic stability.

[166] **ATR Proteins**: Proteins involved in the cellular response to DNA damage and replication stress.

[167] **Proteins Ku70 and Ku80:** Proteins that bind DNA ends, important in non-homologous end joining during DNA repair.

[168] **DNA-PKcs:** A protein kinase essential for the repair of double-strand DNA breaks via non-homologous end joining.

[169] **Ligase IV enzymes:** Enzymes that join broken DNA strands during repair, crucial for maintaining genetic information.

[170] **Ligation:** The process of joining two DNA strands together, an essential step in DNA repair.

[171] **Hybridization:** The process where complementary DNA or RNA strands pair, used in gene mapping and cloning.

Endosymbiotic Gene Transfer (EGT)

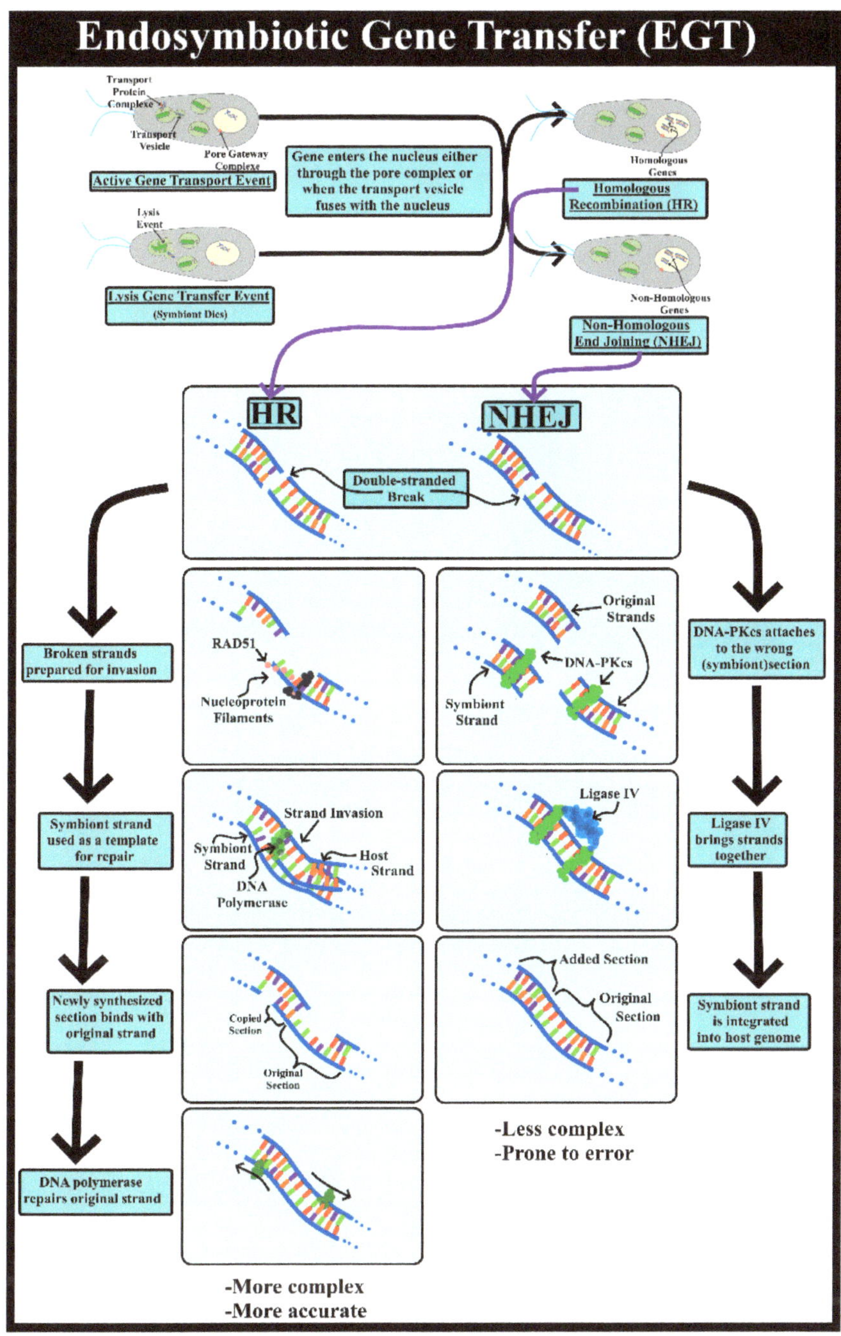

Figure 6-1. EGT By Victor M Rodriguez 2023

6.1.4 Implications for Endosymbiotic Co-Evolution

Once gene integration is complete, the regulatory framework for expression[172] and transcription[173] is updated to ensure that the genes are expressed or suppressed as needed. This process involves the evolution of promoter sequences of DNA[174] so the transcription complex[175] can bind to the gene and execute gene transcription for eventual protein synthesis(Sloan et al., 2018a).

The rapid evolution of an updated regulatory framework is essential for successful genetic integration. This framework must regulate the new genetic material in a mutually beneficial way to ensure the continuation of the endosymbiotic relationship. Systems are developed for metabolic synchronization, enhancing and suppressing specific functions to aid genetic integration. This synchronization allows the symbiont and host to coordinate cellular and metabolic functions, eliminating redundant processes for energy efficiency. RNA interference [176] can further streamline genetic processes by targeting and degrading mRNA[177] strands to prevent protein synthesis if a redundant gene transcribes the mRNA(Sloan et al., 2018a).

---

[172] **Expression**: The process by which a gene's information is used to synthesize functional gene products.

[173] **Transcription**: The synthesis of RNA from a DNA template, the first step in gene expression.

[174] **Promoter sequences of DNA**: DNA regions where RNA polymerase binds to initiate transcription of a gene.

[175] **Transcription complex**: A group of proteins that assemble on DNA to begin transcription.

[176] **RNA interference:** A biological process where RNA molecules inhibit gene expression by destroying specific mRNA molecules.

[177] **mRNA**: Messenger RNA, a type of RNA that conveys genetic information from DNA to the ribosome.

The process of gene suppression and loss of function serves not only to improve efficiency but also to strengthen the endosymbiotic relationship, as many of the lost functions by the symbiont or host can result in the two parties forming an irrevocable dependency.

6.2 Cellular Communication and Reproductive Coordination

Genetic integration and streamlining within an endosymbiotic relationship is only part of a greater biochemical entanglement that occurs when two organisms "decide" to become one. An entire complex apparatus of communication, coordination, and sensing must emerge to ensure the two parties do not become harmful to each other. This apparatus must be able to coordinate everything from metabolism, repairs, maintenance, and even reproduction. These control systems are essential for gene expression and repression by providing the necessary information on the organism's state to express the proper genes(Sloan et al., 2018b).

6.2.1 Calcium Mediated Symbiotic Communication

Calcium is a powerful tool used to control many functions within a cell. In endosymbiosis, calcium can also be integral to nutrient and metabolic regulation. The endoplasmic reticulum (ER)[178] serves as a storage unit of calcium, which is released as needed by the cell for communications. Embedded in the host and symbiont membranes are

---

[178] **Endoplasmic reticulum (ER):** An organelle in cells that plays a role in protein folding and transport.

transport proteins and channels utilized to facilitate the transportation and uptake of nutrients, waste, and other essential molecules. The proteins that compose these channels are sensitive to fluctuations in calcium concentration, which will trigger the transporters and channels to release specific nutrients or inhibit particular channels. Calcium concentration fluctuations and oscillations can code for the release of specific nutrients to a certain quantity(Blackstone, 2015).

When the symbiont lacks a specific nutrient, signaling proteins can be released which interact with the host mechanisms, triggering calcium release into the cytoplasm. This signal is a localized response and can code for nutrients or other processes in a specific location within the cell. When the endosymbiont is no longer nutrient deficient, it can release different messengers, instructing the calcium levels to return to their previous state. This request allows for specific nutrient channels to be deactivated. These calcium-mediated signals can also be synchronized with other pathways to facilitate a coordinated response to the symbiont request(Blackstone, 2015). This may include the activation of ATP synthesis[179] when specific transport protocols require ATP to facilitate the transport of nutrients or proteins.

---

[179] **ATP synthesis:** The biochemical process of producing ATP, the energy currency of the cell.

Calcium Mediated Communication

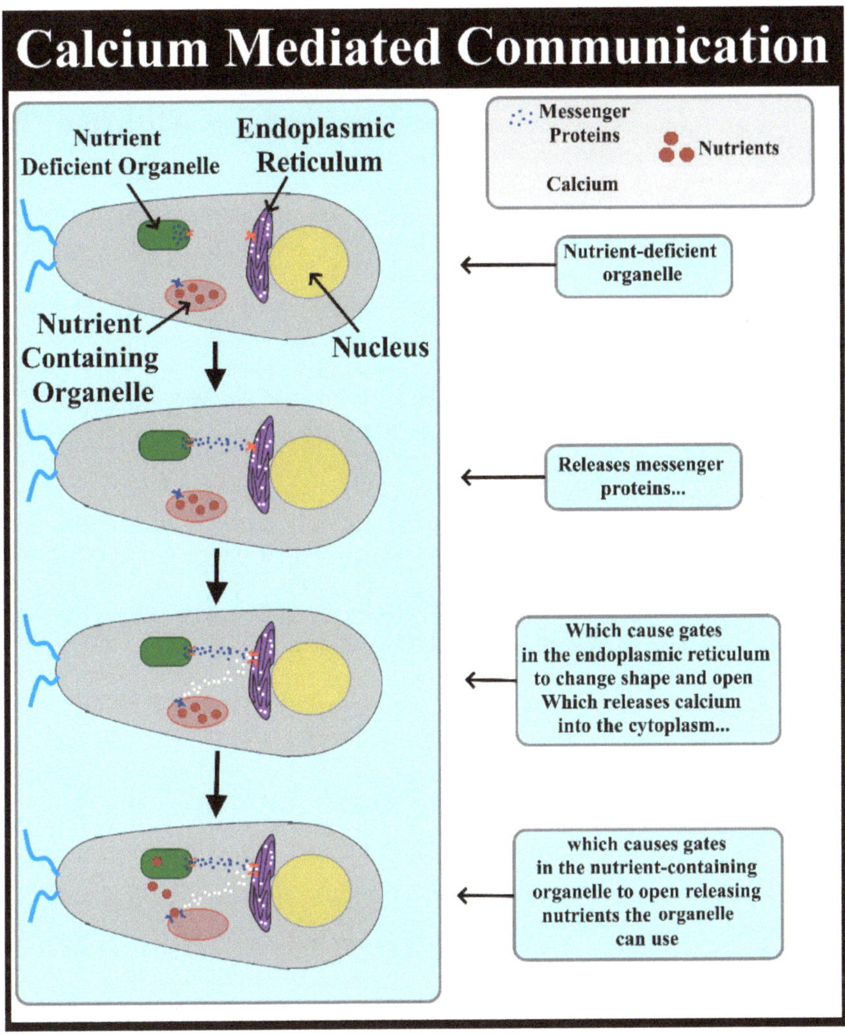

Figure 6-2. Calcium Mediated Communication | By Victor M. Rodriguez

<u>6.2.2 Host Mediated Plastid Division</u>

It is essential to maintain the endosymbiotic relationship so that the host can control the symbiont population within the cell. Improper population controls can lead to resource depletion or overproduction of compounds. As mentioned, horizontal gene transfer transfers many functions from the symbiont to the host nucleus. One of these functions may be reproduction itself. The host mediates the gene for plastid division in fully evolved plastid possessing eukaryotic cells. The tubulin-like FtsZ protein[180] assembles into a ring at the division site of the plastid. This protein is similar to proteins involved in controlling binary fission[181] in cyanobacteria. However, the gene that codes for the production of this protein was transferred from the plastid to the host nucleus at some point in its evolution. Plastid division is a tightly regulated process. This contrasts the endosymbiotic relationships between diatoms of the genus Rhizosolenia, which have endosymbiotic cyanobacteria that still retain the ability to control their cell division(Hashimoto, 2003).

Although plastids can initiate their division, the host encodes and synthesizes the proteins needed to complete the process. Therefore, the plastid division can only occur with the "approval" of the host. Plastid division is usually increased during the lead-up to host cell division. Once an adequate number of plastids are present, the host will ensure that the two resulting daughter cells contain the required number. The host cell's cytoskeleton controls the process of even plastid division. Microtubules

---

[180] **FtsZ protein:** A bacterial protein that initiates cell division, similar to actin in eukaryotes.
[181] **Binary fission:** The process by which prokaryotic cells divide, involving FtsZ protein.

and actin filaments position the plastids in their proper locations just before cell division(Hashimoto, 2003).

Reproduction of Symbionts who have not given up their autonomy can be more like a relationship of give and take. Unlike endosymbiotic relationships between plastids and hosts, symbiotic relationships such as those between *Paramecium* and *Chlorella* require careful regulation of available nutrients. Theoretically, the *Chlorella* living within the *Paramecium* can reproduce at will. The reality is more complicated. Environmental feedback regulates reproduction rates. The cell division rates will increase with nutrient availability and decrease when the waste products produced by the *Chlorella* exceed the ability of the *Paramecium* to expel these products. Unlike with plastid/host cell division, the Cell division of the Paramecium is not as concerned with symbiont equality among the daughter cells. Instead, the distribution of *Chlorella* symbionts is roughly distributed equally by chance. If the two daughter cells receive an unequal amount of symbiont algae cells, the environmental feedback mechanism will tend to regulate the population accordingly. Daughter cells with fewer symbionts will have more nutrients available per symbiont, triggering cell division. Meanwhile, due to stress, the daughter cell with an overabundance of symbiont algae will likely have slower cell division(Xiang et al., 2020).

Interestingly, the paramecium can expel chlorella cells if they start to overcrowd the Paramecium or seek out and acquire more if it lacks enough chlorella to glean any benefit from the relationship.

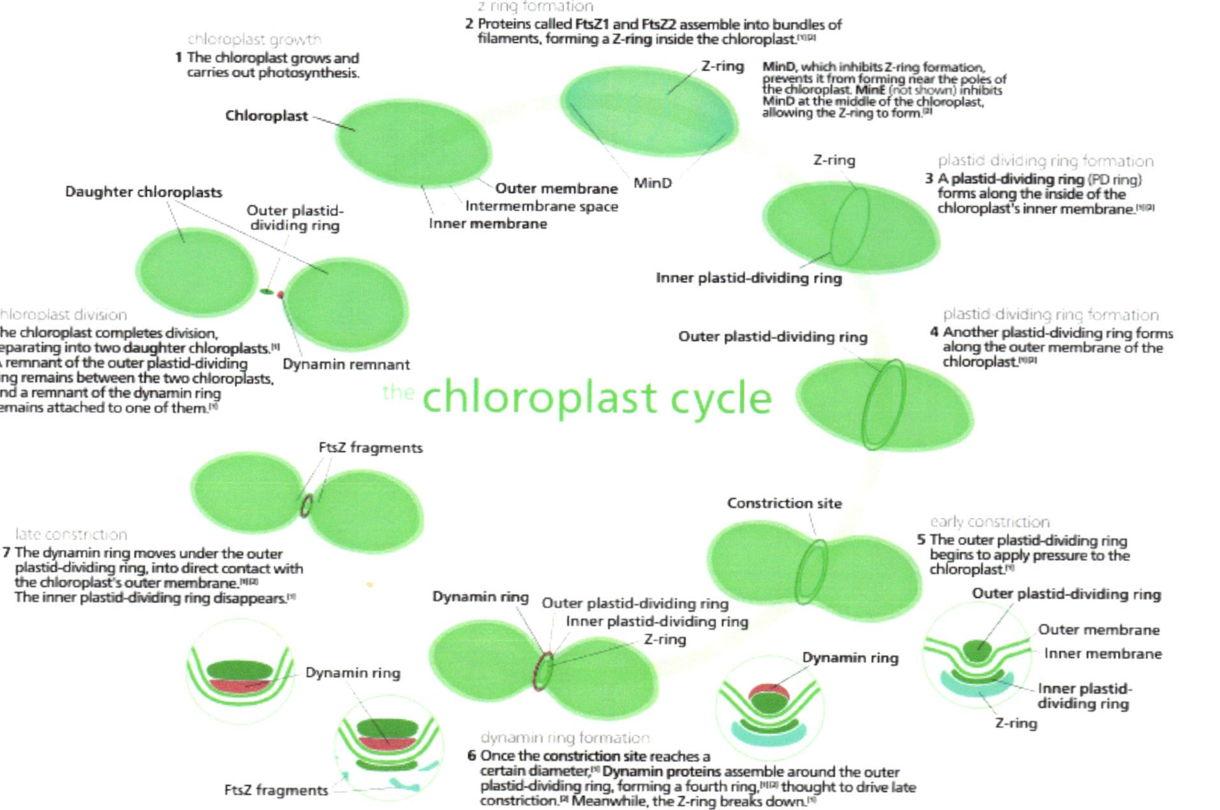

Figure 6-3. Chloroplast Lifecycle | By Kelvinsong - Own work

114

6.3 Chapter Summary

The intricate dance of cellular communication and reproductive coordination underscores the marvel of endosymbiotic relationships in the cellular world. Cells streamline their operations through genetic integration, converging into a more efficient and symbiotic entity. As an essential cellular messenger, calcium bridges the communication gap between the host and symbiont, ensuring seamless coordination and nutrient regulation. Furthermore, the delicate balance of plastid division in the host, guided by processes evolved from ancient partnerships, epitomizes the depth of integration and mutual dependence. This chapter has provided a glimpse into the delicate and complex choreography that ensures the harmonious existence of intertwined life forms. As we continue to unravel the mysteries of cellular communication, one thing becomes clear: life, in all its myriad forms, thrives on collaboration and unity.

Figure 7-1. Sea slug *Elysia chlorotica* (Cai et al., 2019)

116

7. KLEPTOPLASTY

Kleptoplasty is an interes-ting practice within algal and other single and multicellular organisms(Cruz & Cartaxana, 2022a)(Tsuji, 2023). Kleptoplasty is consuming a plastid-containing microbe, then partially digesting the cell, leaving only undigested plastids. The plastid will continue its biochemical metabolic functions as if in its native organism (Cruz & Cartaxana, 2022a). Kleptoplasty has even been observed in animals such as sea slugs and marine flatworms(Van Steenkiste et al., 2019). In the kleptoplastic relationship, the plastid provides the photosynthetic byproducts to its host, as it would its native cell(Cruz & Cartaxana, 2022a). It is important to note that unlike other endosymbiotic relationships mentioned before, kleptoplasty is short-lived, only lasting a few days or months because the host does not provide the nutrients or mechanisms needed to prevent plastid decay(Karnkowska et al., 2023).

The process of kleptoplasty stands out among the many other endosymbiotic events. Most notably, the targeting and uptake are more of a feature than a mistake(Cruz & Cartaxana, 2022b). The host organism has evolved to recognize the plastids by containing them within a protective membrane or through the evolutionary modification of lysosomal mechanisms to avoid plastid breakdown. Meanwhile, the rest of the ingested organism is digested, and the resulting molecules are recycled for use by the host(Pelletreau et al., 2014).

Some dinoflagellates can acquire secondary kleptoplasty by acquiring plastids from ciliates, which, in turn, are obtained from a cryptophyte alga(Tsuji, 2023). The Sea Slug *Elysia chlorotica* is an example which has been studied in depth(Pelletreau et al., 2014). This slug can absorb and retain plastids for months, benefiting from its newly acquired energy source(Laetz & Wägele, 2017). This level of kleptoplasty

Figure 7-2. *Costasiella kuroshimae* aka leaf slug | https://anandamide.substack.com/p/kleptoplastic-conniptions

is a fascinating example in that slugs are multicellular organisms. The complexity of such a function is an interesting question in biology. Although the exact evolutionary journey of sea slugs and other kleptoplastic multicellular organisms, such as flatworms, is poorly understood, there are several possible modalities for kleptoplasty development(Van Steenkiste et al., 2019). These relationships may have started as symbiotic relationships that gradually became one-sided. There may also have been selective pressures resulting from nutrient-scarce environments. By evolutionary chance, the organisms that evolved an inability to digest algal plastids benefitted from what would otherwise have been a disadvantage(Karnkowska et al., 2023).

There are no confirmed examples of kleptoplasty resulting in a permanent relationship. Whether in single or multicellular hosts, kleptoplasty is, as far as is known, a temporary arrangement. The reason for this is that the biological steps that would need to evolve from kleptoplasty to plastid integration may be unsurmountable. With traditional endosymbiotic relationships, enough of the original plastid owner's biochemical support system remains long enough for the required co-evolution and gene transfer to allow for proper integration. In kleptoplasty, this gene transfer, membrane modification, and co-evolution would have to occur within a few months at most before the plastid decays from lack of maintenance(Karnkowska et al., 2023).

8. SUSTAINING AND TERMINATING ENDOSYMBIOTIC RELATIONSHIPS

8.1 Autophagy and Endosymbiotic Maintenance

In an endosymbiotic relationship, a symbiont or organelle may become detrimental to the host's survival. In this case, cells have evolved a process known as autophagy, in which a cell detects some error in the plastid or symbiont. The host creates a double-membraned vesicle called an autophagosome around the target organelle or symbiont, and digestion and recycling begins(Tooze et al., 2014). The autophagosome fuses with lysosomes to start the degradation process(Yamamoto et al., 2023). It is important to note that this process is not unique to endosymbiotic relationships. Autophagy is employed as a cellular maintenance mechanism whenever a component is no longer needed or if its continued existence is detrimental to the function of the entire cell(Khandia et al., 2019). When autophagy is used to target a foreign symbiont that is not fully integrated into the biome of the host, it is called xenophagy. This process is used for intruders such as viruses and bacteria as well(Cong et al., 2020).

Cells have various systems to check the health of their organelles and symbiotic entities. Autophagy is a cellular degradation process initiated when proteins misfold and are marked for breakdown by a protein called ubiquitin. The mechanistic Target of Rapamycin (mTOR) [182] pathway regulates the cell's nutrient status. When nutrients are plentiful, mTOR is active, promoting protein synthesis and the creation of other

---

[182] **mTOR**: A central cell growth regulator responding to nutrients, growth factors, and energy levels.

molecular components. Conversely, when nutrients are in short supply, mTOR is deactivated. This deactivation includes the mTOR Complex 1 (mTORC1)[183], which suppresses the activation of the ULK1 kinase[184], a trigger for autophagy. However, when mTORC1 is inactive, ULK1, alongside other kinases and complexes, can initiate autophagy. The cell also has an energy/nutrient sensor, the AMP-activated protein kinase (AMPK)[185], which responds to low energy levels. Once energy falls below a certain threshold, AMPK stimulates the Tuberous Sclerosis Proteins (TSC1/2)[186], leading to the inactivation of mTORC1(McKee-Muir & Russell, 2017).

8.2 Endosymbiosis termination

A host cell diligently oversees its endosymbiotic relationships to maintain mutual benefits. However, if plastids or external organisms cease to benefit the host, they may need to be discarded. Sometimes, the host must end the endosymbiotic relationship entirely. When the symbiont is not fully integrated into the host's biological processes, the cell has alternatives beyond autophagy to manage such situations.

In the case of *Symbiodinium* algae's relationship with coral, when temperatures cause cellular stress, the coral will use a process similar to

---

[183] **mTORC1**: A protein complex that modulates cell growth, protein synthesis, and autophagy based on cellular conditions.
[184] **ULK1 kinase**: An enzyme that initiates autophagy when nutrients are scarce and mTOR activity is low.
[185] **AMPK**: An energy sensor that activates energy-producing pathways when cellular energy is depleted.
[186] **TSC1/2**: Proteins that inhibit mTORC1, acting as tumor suppressors and regulators of cell growth.

exocytosis [187] to expel the symbiotic algae, causing coral bleaching. Dinoflagellates also utilize this process to expel an algal symbiont if there is a buildup of reactive oxygen species or other conditions adverse to the continuation of the symbiotic relationship(Imanian et al., 2012).

The process may differ depending on the organism. However, exocytosis involves the recognition of the symbiont as either damaged or no longer metabolically beneficial. The host's immune response system may mediate this recognition process. Once the organism is recognized and marked for expulsion, a membrane-bound vesicle encapsulates the organism, and the vesicle fuses with the membrane. The organism is then ejected similarly to waste products expelled from a cell.

<u>8.3 Chapter Summary</u>

In summary, the maintenance and potential termination of endosymbiotic relationships are crucial processes that ensure the survival and health of host organisms. Autophagy is a selective and regulated means of removing malfunctioning or superfluous endosymbionts or organelles, thereby maintaining cellular homeostasis. When autophagy targets non-native entities within the cell, the process is referred to as xenophagy, showcasing a defense mechanism against potential pathogens.

Furthermore, when conditions within the host organism shift unfavorably, leading to the detriment of the symbiotic balance, termination mechanisms such as those observed in coral-dinoflagellate relationships come into play. The host can actively expel symbionts

---

[187]**Exocytosis**: A process where cells expel molecules through vesicle fusion with the plasma membrane.

through processes akin to exocytosis, a clear indication of the dynamic and responsive nature of endosymbiotic relationships.

Monitoring systems within the host, such as the mTOR pathway, continuously assess the state of symbionts, marking for degradation or expulsion those that no longer contribute positively to the host's well-being. These pathways highlight an intricate interplay between nutrient availability, energy status, and symbiont viability, underpinning the adaptive responses to internal and external environmental challenges.

As we delve deeper into the complexities of these relationships, it becomes evident that the line between maintaining and terminating endosymbiotic partnerships is finely tuned. Host organisms must constantly negotiate these boundaries to preserve their health and ensure evolutionary success. Future research may shed light on additional mechanisms and controls that govern these fascinating cellular interactions, offering insights into the delicate balance of living in unity and the decisive moments when separation becomes the key to survival.

9. DISCUSSION

As our exploration of algal evolution comes to a close, it becomes clear that algae and cyanobacteria are not merely passive bystanders in the story of life but are their foundational architects. From the primordial chaos of Earth's early days, these photosynthetic pioneers crafted an atmosphere suitable for life as we know it, seeding the planet with the breath of life—oxygen.

Our journey began by unraveling the complex evolutionary lineage of algae. It is a narrative not of straight lines but an interconnected web where the cooperative march of endosymbiosis is a recurring theme. These ancient partnerships forged the diversity of life, a process steered by symbiosis, endosymbiosis, and lateral gene transfer.

In the story of life on Earth, algae and cyanobacteria are the unsung heroes, the foundations of our global ecosystem. Through their relentless photosynthetic drive, Earth transitioned from a barren rock to a blue-green oasis teeming with life. The ancient symbiotic embrace between a cyanobacterium and a primordial eukaryotic host gave birth to the chloroplast, the powerhouses of photosynthesis that now reside in every green leaf and algal cell.

This book has chronicled the evolutionary genius of algae, from their role in oxygenating the atmosphere to their knack for symbiotic relationships that drive the evolution of other life forms. Their propensity for such relationships has altered the story of life, fostering biodiversity and ecological resilience. As primary producers, algae and cyanobacteria are the bedrock upon which the global ecosystem stands; their ability to use sunlight to produce nutrients supports the complex food webs that sustain marine and terrestrial life.

Figure 9-1. Illustration of the Cambrian period ocean, Credit: ©iStock

One cannot overstate the ecological significance of algae. They continue to oxygenate our planet, regulate carbon cycles, and support many life forms. From the tiniest zooplankton to the largest whales, life in the oceans and on land is intertwined with the fate of algae. On land, their distant cousins, plants, have colonized every conceivable niche, with their algal ancestors' legacy evident in the green pigment that colors our world.

However, the future presents us with threats and opportunities. Climate change looms large, its multifaceted impacts poised to disrupt the delicate balance these ancient organisms maintain. Just as cyanobacteria reshaped the entire global ecosystem, as a species we stand to completely reshape the future of life on earth. Rising temperatures, acidifying oceans, and altered aquatic nutrient cycles spell unprecedented stress for algal communities. Yet, in adversity lies opportunity—algae have shown remarkable resilience and adaptability through Earth's tumultuous past, suggesting a potential source of hope. Their ability to adapt to changing conditions to form new symbiotic alliances may once again play a pivotal role in sustaining and even reshaping ecosystems in the face of anthropogenic climate change.

The story of algae is a story of life itself—of its origins, its evolution, and its persistence. It is a story that underscores the importance of preserving the delicate balance of our planet's ecosystems, a reminder that the welfare of the smallest microorganisms is inextricably linked to the health of the entire biosphere. As we look to the future, we must carry forward the lessons from these microscopic marvels. Their history is our history, their future, our own.

As we venture forth, we must pause to acknowledge the remarkable evolutionary innovations of algae that have made Earth habitable. Their early domination of aquatic environments and subsequent colonization of land laid down the biological and geochemical foundation for all subsequent life forms. The proliferation of algae during the Precambrian and Phanerozoic eons was a period of extensive biochemical innovation, which saw the atmosphere's oxygen levels rise and dip in a complex dance with life's demands.

These photosynthetic powerhouses have crafted our atmosphere and actively shaped the course of evolution through endosymbiosis. The evidence presented in this book elucidates how primary and secondary

125

endosymbiotic events have peppered the tree of life, creating a mosaic of organisms that defy traditional taxonomic boundaries. This genetic exchange has been pivotal in driving the evolutionary success of eukaryotic lineages, with algae often at the heart of these transformations.

The concept of symbiosis extends beyond the cellular level, influencing entire ecosystems. Algae's ability to form partnerships ranges from mutualistic to parasitic, each interaction playing a role in ecological stability and the adaptive potential of species. The resilience of coral reefs, the productivity of wetlands, and the balance of marine food chains all testify to algae's ecological roles within their communities.

Figure 9-3. 2009 bloom on Lake Erie | Credit: NOAA Great Lakes Environmental Research Laboratory

In modern times, algae continue to be a cornerstone of ecological cycles. They are the unsung heroes in the fight against climate change, sequestering vast amounts of carbon dioxide through photosynthesis. Yet, their roles are not restricted to carbon fixation. Algal blooms, for instance, are a natural phenomenon with the dual potential to support life or to suffocate it, depending on the environmental context. The dual nature of algae as both life-givers and, at times, agents of destruction encapsulates the delicate balance of ecological systems.

The threats posed by climate change are multifaceted, impacting algae in ways that are still not fully understood. Ocean acidification, a byproduct of increased carbon emissions, threatens the very fabric of marine ecosystems by disrupting the calcification processes critical to coral and phytoplankton. Warmer waters may shift algal distributions, altering the timing and intensity of blooms, with cascading effects up the food web. Yet, within these threats, there are also opportunities for algae to demonstrate their adaptive capabilities, potentially offering biotechnological solutions to some of our most pressing environmental challenges.

As we look to a future marked by uncertainty, it is clear that algae will play a central role in shaping our planet's response to climate change. Their evolutionary history is replete with examples of adaptation and resilience. The burgeoning field of algal biotechnology is exploring innovative ways to harness this resilience, using algae for biofuel production, bioremediation, and as a sustainable food source. These applications reflect algae's versatility and offer a glimpse into a future where humanity works in concert with these primordial organisms to create a more sustainable and resilient biosphere.

Algae, the ancient stewards of Earth, offer us a blueprint for resilience and adaptability. We must heed their ecological lessons, recognizing our role not as conquerors of nature but as participants in a complex and interdependent web of life. The future of our global ecosystem may well hinge on our ability to understand and emulate the collaborative and adaptive spirit of algae as we forge a sustainable path forward in an ever-changing world.

As a species, we must confront the effects of climate change on algal ecosystems, entities that are not merely participants in but central to Earth's biological machinery. As the Earth's thermostat climbs, the biological responses of algae are complex and far-reaching. Rising temperatures catalyze change, instigating a cascade of biological and ecological transformations within algal communities. Warm waters may accelerate metabolic rates, leading to increased algal growth in some species, but this rapid proliferation can disrupt the delicate equilibrium of aquatic ecosystems. While natural, the phenomenon of algal blooms can thus be exacerbated to the point of creating hypoxic conditions that suffocate marine life, a state known as eutrophication.

As CO_2 levels increase, oceans absorb the excess, acidifying the ocean. This change in pH can impair calcifying algae, such as *coccolithophores*, whose calcium carbonate structures are fundamental to their survival and to the marine carbon cycle. The dissolution of these microscopic organisms can reduce the ocean's capacity to sequester carbon, further exacerbating global warming.

Figure 9-4. Marine dead zone | Sirachai Arunrugstichai / Getty Images

Furthermore, climate change is poised to rewrite the geographic narratives of algal species. As waters warm, algal populations shift towards the poles, following the ocean temperatures they adapt to. This migration can lead to the establishment of algae species in new habitats, disrupting already established ecosystems. The loss of algae from their native ranges can also have dramatic effects on local species that rely on them, from the tiniest of zooplankton to the largest marine mammals, altering the stability of food webs.

Alterations in the timing and intensity of algal blooms due to climate change have profoundly affected the seasonal cycles of biological activities. Mismatches in the timing of algal blooms can disrupt the life cycles of dependent species, such as fish and birds, which synchronize their reproductive cycles with these blooms for food availability. Such mismatches can reduce reproductive success and survival rates, with potential knock-on effects throughout the ecosystem.

In addition to temperature and acidity, the changing climate affects the nutrient dynamics in aquatic systems. Melting ice caps and glaciers, altered river flows and increased stratification of water bodies change the distribution and concentration of nutrients such as nitrogen and phosphorus, vital for algal growth. Some algal species may thrive under these new conditions, while others may struggle to survive, leading to changes in community composition and potentially losing biodiversity. However, in the face of these challenges, algae's true potential comes to the fore. Their evolutionary history is marked by resilience and adaptability—traits now being harnessed in the fight against climate change. Algal biotechnology presents a beacon of hope, offering sustainable solutions in biofuel production, carbon sequestration, and waste water treatment. These innovative applications not only leverage algae's inherent capabilities but also embody a shift towards a more symbiotic relationship with our environment.

Figure 9-5. Credit Chapman University

The future of algae in a changing world is inextricably linked to our actions and choices. As we grapple with the realities of climate change, we must recognize the integral role that these microscopic organisms play in our planet's health. Protecting and understanding algal systems is not just an environmental imperative but a necessity for our continued existence.

In conclusion, the story of algae is a testament to the power of life to adapt, evolve, and flourish under the most challenging conditions. It is a narrative that underscores the interconnectedness of all life on Earth and the delicate balance that sustains it. As we move forward, let us draw inspiration from the resilience and adaptability of algae, embracing our role as stewards of this incredible planet.

Figure 9-6. Credit Cesar Torres

REFERENCES

A. Reid, F. B. (2021). A review on diatom biosilicification and their adaptive ability to uptake other metals into their frustules for potential application in bone repair. *Royal Society of Chemistry*, 6728-6737.

Abas, A. (2021). A systematic review on biomonitoring using lichen as the biological indicator: A decade of practices, progress and challenges. *Ecological Indicators*.

Alcott, L. J., Mills, B. J. W., Bekker, A., & Poulton, S. W. (2022). Earth's Great Oxidation Event facilitated by the rise of sedimentary phosphorus recycling. *Nature Geoscience*, *15*(3), 210–215. https://doi.org/10.1038/s41561-022-00906-5

Alexei Y. Kostygov, A. K. (2021). Euglenozoa: taxonomy, diversity and ecology, symbioses and viruses. *Open Biology*.

Alle A. Y. Lie, Z. L. (2017). Effect of light and prey availability on gene expression of the mixotrophic chrysophyte, Ochromonas sp. *BMC Genomics*.

Alle A. Y. Lie, Z. L. (2018). A tale of two mixotrophic chrysophytes: Insights into the metabolisms of two Ochromonas species (Chrysophyceae) through a comparison of gene expression. *PLOS ONE* .

Alleva, K., Chara, O., & Amodeo, G. (2012). Aquaporins: Another piece in the osmotic puzzle. *FEBS Letters*, *586*(19), 2991–2999. https://doi.org/10.1016/j.febslet.2012.06.013

Amna Komal Khan, 1. H. (2020). An Insight into the Algal Evolution and Genomics. *Biomolecules*.

Andersson, J. O. (2009). Horizontal Gene Transfer Between Microbial Eukaryotes. In M. G. Gogarten, *Horizontal Gene Transfer. Methods in Molecular Biology* (pp. 473-487). Springer.

András Szilágyi, P. S. (2020). Phenotypes to remember: Evolutionary developmental memory capacity and robustness. *PLos Computational Biology*.

Andrew F. Torres, D. A.-G. (2021). Zooxanthellae Diversity and Coral-Symbiont Associations in the Philippine Archipelago: Specificity and Adaptability Across Thermal Gradients. *Frontiers in Microbiology*.

Andrew T. Canada, E. J. (1989). Superoxide dismutase: its role in xenobiotic detoxification. *Pharmacology & Therapeutics*, 285-295.

Anne-Marie Pauwels, M. T. (2017). Patterns, Receptors, and Signals: Regulation of Phagosome Maturation. *Trends in Immunology*, 407-422.

AquaPortail. (2023, September 1). *Apicomplex*. Retrieved from AquaPortail: https://www.aquaportail.com/dictionnaire/definition/14015/apico mplexe

Archibald, J. M. (2015). Genomic perspectives on the birth and spread of plastids. *Proceedings of the National Academy of Sciences*, *112*(33), 10147–10153. https://doi.org/10.1073/pnas.1421374112

Artem A. Trofimov, A. A. (2019). Deep data analytics for genetic engineering of diatoms linking genotype to phenotype via machine learning. *npj Computational Materials*.

Arwa Elaagip, S. A. (2022). Apicoplast Dynamics During Plasmodium Cell Cycle. *Frontiers in Cellular and Infection Microbiology.*

Bahareh Nowruzi, N. B. (2021). Plant-cyanobacteria interactions: Beneficial and harmful effects of cyanobacterial bioactive compounds on soil-plant systems and subsequent risk to animal and human health. *Phytochemistry.*

Beatriz Díaz-Reinoso, I. R.-G. (2021). Towards greener approaches in the extraction of bioactives from lichens. *Reviews in Environmental Science and Bio/Technology,* 917–942.

Behzad Imanian, J.-F. P. (2012). Tertiary Endosymbiosis in Two Dinotoms Has Generated Little Change in the Mitochondrial Genomes of Their Dinoflagellate Hosts and Diatom Endosymbionts. *PLoS One.*

Behzad Imanian, P. J. (2007). The dinoflagellates Durinskia baltica and Kryptoperidinium foliaceum retain functionally overlapping mitochondria from two evolutionarily distinct lineages. *BMC Ecology and Evolution.*

Bennici, A. (2008). Origin and early evolution of land plants. *Communicative & Integrative Biology,* 212-218.

Bettina E. Schirrmeister, M. G. (2015). Cyanobacteria and the Great Oxidation Event: evidence from genes and fossils. *Palaeontology,* 769-785.

Bickford, M. E. (2013). Role of geobiology in the astrobiological exploration of the Solar System. In J. D. Farmer, *The Web of Geological Sciences: Advances, Impacts, and Interactions* (pp. 567-589). Geological Society of America.

Biography, C. D. (2023, November 15). *De Bary, (Heinrich) Anton.* Retrieved from Encyclopedia.com:

https://www.encyclopedia.com/science/dictionaries-thesauruses-pictures-and-press-releases/de-bary-heinrich-anton

Blackstone, N. W. (2015). The impact of mitochondrial endosymbiosis on the evolution of calcium signaling. Cell Calcium, 57(3), 133–139. https://doi.org/10.1016/j.ceca.2014.11.006

Blackwell, W. H. (2009). CHROMISTA REVISITED: A DILEMMA OF OVERLAPPING. *Phytologia*, 191-225.

Bonilla, B., Hengel, S. R., Grundy, M. K., & Bernstein, K. A. (2020). RAD51 Gene Family Structure and Function. Annual Review of Genetics, 54(1), 25–46. https://doi.org/10.1146/annurev-genet-021920-092410

Bożena Zakryś 1, R. M. (2017). Evolutionary Origin of Euglena. In S. S. Schwartzbach, *Advances in Experimental Medicine and Biology*. Springer, Cham.

Brenda S. Prattel, T. T. (2021). Comparative genomic insights into culturable symbiotic cyanobacteria from the water fern Azolla. *Microbial Genomics*.

Bruce A. Curtis, G. T. (2012). Algal genomes reveal evolutionary mosaicism and the fate of nucleomorphs. *Nature*, 59-65.

Burriesci, M. S., Raab, T. K., & Pringle, J. R. (2012). Evidence that glucose is the major transferred metabolite in dinoflagellate–cnidarian symbiosis. Journal of Experimental Biology, 215(19), 3467–3477. https://doi.org/10.1242/jeb.070946

Cai, H., Li, Q., Fang, X., Li, J., Curtis, N. E., Altenburger, A., Shibata, T., Feng, M., Maeda, T., Schwartz, J. A., Shigenobu, S., Lundholm, N., Nishiyama, T., Yang, H., Hasebe, M., Li, S., Pierce, S. K., & Wang, J. (2019). A draft genome assembly of the solar-powered sea slug Elysia chlorotica. Scientific Data, 6(1), 190022. https://doi.org/10.1038/sdata.2019.22

Cardona, T. (2016). Origin of Bacteriochlorophyll a and the Early Diversification of Photosynthesis. *PLOS ONE*

Carmen M. Cromer, L. M. (2021). Mechanoreceptors. In J. S. Vonk, *Encyclopedia of Animal Cognition and Behavior* (pp. 1-4). Springer.

Catherine F. Demoulin, Y. J. (2019). Cyanobacteria evolution: Insight from the fossil record. *Free Radical Biology and Medicine*, 206-223.

Catherine Kistner, M. P. (2002). Evolution of signal transduction in intracellular symbiosis. *Trends in Plant Science*, 511-518.

Cavalier-Smith, T. (2018). Kingdom Chromista and its eight phyla: a new synthesis emphasising periplastid protein targeting, cytoskeletal and periplastid evolution, and ancient divergences. *Protoplasma*, 297-357.

Cédric Hubas, B. J. (2023). Proliferation of Purple Sulphur Bacteria at the Sediment Surface Affects Intertidal Mat Diversity and Functionality. *PlosOne*.

Chakraborty, A. (2016). Understanding the biology of the Plasmodium falciparum apicoplast; an excellent target for antimalarial drug development. *Life Sciences*, 104-110.

Chris Greening, T. L. (2020). Formation and function of bacterial organelles. *Nature Reviews Microbiology*, 677-689.

Christa Decker, S. S. (2021). Pro-Resolving Ligands Orchestrate Phagocytosis. *Frontiers in Immunology*.

Christoph Kneip, C. V. (2008). The cyanobacterial endosymbiont of the unicellular algae Rhopalodia gibba shows reductive genome evolution. *BMC Evolutionary Biology*.

Christopher Jackson, A. H. (2018). Plastid phylogenomics with broad taxon sampling further elucidates the distinct evolutionary origins and timing of secondary green plastids. *Scientific Reports*.

Claudia Carranza MSc, L. C.-G. (2019). Several Routes to the Same Destination: Inhibition of Phagosome-Lysosome Fusion by Mycobacterium tuberculosis. *The American Journal of the Medical Sciences*, 184-194.

Cong, Y., Dinesh Kumar, N., Mauthe, M., Verlhac, P., & Reggiori, F. (2020). Manipulation of selective macroautophagy by pathogens at a glance. Journal of Cell Science, 133(10). https://doi.org/10.1242/jcs.240440

Consolación Álvarez, L. J.-R.-P.-F.-H. (2023). Symbiosis between cyanobacteria and plants: from molecular studies to agronomic applications. *Journal of Experimental Botany*, 6145–6157.

Cournoyer, J. E., Altman, S. D., Gao, Y., Wallace, C. L., Zhang, D., Lo, G.-H., Haskin, N. T., & Mehta, A. P. (2022). Engineering artificial photosynthetic life-forms through endosymbiosis. Nature Communications, 13(1), 2254. https://doi.org/10.1038/s41467-022-29961-7

Cruz, S., & Cartaxana, P. (2022a). Kleptoplasty: Getting away with stolen chloroplasts. PLOS Biology, 20(11), e3001857. https://doi.org/10.1371/journal.pbio.3001857

Cruz, S., & Cartaxana, P. (2022b). Kleptoplasty: Getting away with stolen chloroplasts. PLOS Biology, 20(11), e3001857. https://doi.org/10.1371/journal.pbio.3001857

Dahan, R. A., Duncan, R. P., Wilson, A. C., & Dávalos, L. M. (2015). Amino acid transporter expansions associated with the evolution of obligate endosymbiosis in sap-feeding insects (Hemiptera: sternorrhyncha). *BMC Evolutionary Biology*, 15(1), 52. https://doi.org/10.1186/s12862-015-0315-3

Dales, R. P. (1960). On the pigments of the Chrysophyceae. *Journal of the Marine Biological Association of the United Kingdom*, 693-699.

Daniel P. Glavin, J. L. (2004). Survival of Amino Acids in Micrometeorites During Atmospheric Entry. *Astrobiology*.

David G. Adams, P. S. (2008). Cyanobacteria–bryophyte symbioses. *Journal of Experimental Botany*, 1047–1058.

David G. Adams, P. S. (2012). Cyanobacterial Symbioses. In B. Whitton, *Ecology of Cyanobacteria II* (pp. 593–647). Dordrecht: Springer.

David L. Hawksworth, M. G. (2020). Lichens redefined as complex ecosystems. *New Phytologist*, 1362–1375.

Dean, A. D., Minter, E. J. A., Sørensen, M. E. S., Lowe, C. D., Cameron, D. D., Brockhurst, M. A., & Jamie Wood, A. (2016). Host control and nutrient trading in a photosynthetic symbiosis. Journal of Theoretical Biology, 405, 82–93. https://doi.org/10.1016/j.jtbi.2016.02.021

Debashish Bhattacharya, H. S. (2003). Photosynthetic eukaryotes unite: endosymbiosis connects the dots. *BioEssays*.

Debashish Bhattacharya, T. N. (2008). ENDOSYMBIOTIC AND HORIZONTAL GENE TRANSFER IN CHROMALVEOLATES. *Journal of Phycology*, 7-10.

Deepti Dabral, J. R. (2017). Phospholipase A2: Potential roles in native membrane fusion. *The International Journal of Biochemistry & Cell Biology*, 1-5.

Eileen Uribe-Querol, C. R. (2020). Phagocytosis: Our Current Understanding of a Universal Biological Process. *Frontiers in Immunology*.

El Mahdi Bendif, I. P. (2023). Rapid diversification underlying the global dominance of a cosmopolitan phytoplankton. *IAME J*, 630-640.

Elisabeth Hehenberger, B. I. (2014). Evidence for the Retention of Two Evolutionary Distinct Plastids in Dinoflagellates with Diatom Endosymbionts. *Genome Biology and Evolution*, 2321-2334.

Ellen Yeh, J. L. (2011). Chemical Rescue of Malaria Parasites Lacking an Apicoplast Defines Organelle Function in Blood-Stage Plasmodium falciparum. *PLoS Biology*.

Encyclopaedia, T. E. (2023). golden algae. In T. E. Britannica, *Britannica.* Encyclopedia Britannica.

Endosymbiosis and the Evolution of Eukaryotes. (n.d.). Retrieved from libretexts.org:
https://bio.libretexts.org/Bookshelves/Introductory_and_General
_Biology/Book%3A_General_Biology_(Boundless)/23%3A_Pro
tists/23.01%3A_Eukaryotic_Origins/23.1C%3A_Endosymbiosis
_and_the_Evolution_of_Eukaryotes#:~:text=Figure%20,already
%20had%20a%20double%20me

Fangru Nan, J. F. (2017). Origin and evolutionary history of freshwater Rhodophyta: further insights based on phylogenomic evidence. *Scientific Reports*.

Filip Husnik, J. P. (2017). Functional horizontal gene transfer from bacteria to eukaryotes. *Nature Reviews Microbiology*, 67-79.

Finan, T. M. (2002). Evolving Insights: Symbiosis Islands and Horizontal Gene Transfer. *Journal of Bacteriology*, 2855-2856.

Fitzgeorge-Balfour, T. (2021, July 15). When corals meet algae: first stages of symbiosis seen for the first time. *Frontiers in Science News*.

Fitzgerald, D. M., & Rosenberg, S. M. (2019). What is mutation? A chapter in the series: How microbes "jeopardize" the modern synthesis. PLOS Genetics, 15(4), e1007995. https://doi.org/10.1371/journal.pgen.1007995

Flügge, U.-I., Westhoff, P., & Leister, D. (2016). Recent advances in understanding photosynthesis. F1000Research, 5, 2890. https://doi.org/10.12688/f1000research.9744.1

Francisco Figueroa-Martinez, C. J.-P. (2018). Plastid Genomes from Diverse Glaucophyte Genera Reveal a Largely Conserved Gene Content and Limited Architectural Diversity. *Genome Biology Evolution*, 174-188.

François Thomas, A. C. (2014). Kelps feature systemic defense responses: insights into the evolution of innate immunity in multicellular. *New Phytologist*, 567-576.

Frederik Leliaert, D. R. (2012). Phylogeny and Molecular Evolution of the Green Algae. *Critical Reviews in Plant Sciences*, 1-46.

Gabr, A., Grossman, A. R., & Bhattacharya, D. (2020). Paulinella , a model for understanding plastid primary endosymbiosis. Journal of Phycology, 56(4), 837–843. https://doi.org/10.1111/jpy.13003

Gang Fu, C. N. (2014). Proteomics analysis of heterogeneous flagella in brown algae (stramenopiles). *Protist*, 662-675.

Garrett, R. A. (2014). A backward view from 16S rRNA to archaea to the universal tree of life to progenotes: reminiscences of Carl Woese. *RNA Biology*, 232-235.

Geoffrey Ian McFadden, E. Y. (2017). The apicoplast: now you see it, now you don't. *International Journal for Parasitology*, 137-144.

Giada Tortorelli, C. R. (2022). Cell surface carbohydrates of symbiotic dinoflagellates and their role in the establishment of cnidarian–dinoflagellate symbiosis. *The ISME Journal*, 190-199.

Goro Tanifuji, R. K. (2020). Comparative Plastid Genomics of Cryptomonas Species Reveals Fine-Scale Genomic Responses to Loss of Photosynthesis. *Genome Biology and Evolution*, 3926–3937.

Gray, M. W. (2017). Lynn Margulis and the endosymbiont hypothesis: 50 years later. *Mol Biol Cell*, 1285-1287.

Gregory, T. R. (2008). Evolution as Fact, Theory, and Path. Evolution: Education and Outreach, 1(1), 46–52. https://doi.org/10.1007/s12052-007-0001-z

Gross, M. (2012). The mysteries of the diatoms. *Current Biology*, 581-585.

Hans Kuhn, J. W. (1981). Molecular Self-Organization and the Origin of Life. *Versammlung der Gesellschaft Deutscher Naturforscher und Ärzte.*

Harry W. Rathbone, K. A. (2021). Scaffolding proteins guide the evolution of algal light harvesting antennas. *Nature Communications.*

Hashimoto, H. (2003). Plastid division: Its origins and evolution (pp. 63–98). https://doi.org/10.1016/S0074-7696(02)22012-4

Hisashi Endo, H. O. (2018). Contrasting biogeography and diversity patterns between diatoms and haptophytes in the central Pacific Ocean. *Scientific Reports*.

Honegger, R. (2000). Simon Schwendener (1829–1919) and the Dual Hypothesis of Lichens. *The Bryologist*, 307-313.

Huanle Liu, T. G.-P. (2018). Symbiodinium genomes reveal adaptive evolution of functions related to coral-dinoflagellate symbiosis. *Communications Biology*.

Hwan Su Yoon, J. D. (2002). The single, ancient origin of chromist plastids. *Proceedings of the National Academy of Sciences*, 15507–15512.

I. N. Stadnichuk, V. V. (2021). Endosymbiotic Origin of Chloroplasts in Plant Cells' Evolution. *Russian Journal of Plant Physiology* , 1-16.

Ilan Chet, S. F. (1971). Chemical Detection of Microbial Prey by Bacterial Predators. *Ilan Chet, Sam Fogel, Ralph Mitchell*, 863–867.

Imanian, B., Pombert, J.-F., Dorrell, R. G., Burki, F., & Keeling, P. J. (2012). Tertiary Endosymbiosis in Two Dinotoms Has Generated Little Change in the Mitochondrial Genomes of Their Dinoflagellate Hosts and Diatom Endosymbionts. PLoS ONE, 7(8), e43763. https://doi.org/10.1371/journal.pone.0043763

Iwai, S., Fujita, K., Takanishi, Y., & Fukushi, K. (2019). Photosynthetic Endosymbionts Benefit from Host's Phagotrophy, Including Predation on Potential Competitors. Current Biology, 29(18), 3114-3119.e3. https://doi.org/10.1016/j.cub.2019.07.074

Ines A. Aschenbrenner, T. C. (2016). Understanding Microbial Multi-Species Symbioses. *Frontiers in Microbiology*.

Jan Janouskovec, A. H. (2010). A common red algal origin of the apicomplexan, dinoflagellate, and heterokont plastids. *Proceedings of the National Academy of Science*, 10949-10954.

Jason E. Cournoyer, S. D.-l.-H. (2022). Engineering artificial photosynthetic life-forms through endosymbiosis. *Nature Communications* .

Jason M. Kinchen, K. S. (2008). Phagosome maturation: going through the acid test. *Nature Reviews Molecular Cell Biology*, 781-795.

Jean-Baptiste Raina, V. F. (2019). The role of microbial motility and chemotaxis in symbiosis. *Nature Reviews Microbiology*, 284-294.

Jean-François Pombert, E. R. (2012). Evidence for Transitional Stages in the Evolution of Euglenid Group II Introns and Twintrons in the Monomorphina aenigmatica Plastid Genome. *PLOS ONE*.

Jeffrey D Leblond, A. D. (2012). Sterols of the green-pigmented, aberrant plastid dinoflagellate, Lepidodinium chlorophorum (Dinophyceae). *Protist*, 38-46.

Jenny A. Nguyen, R. M. (2021). Better Together: Current Insights Into Phagosome-Lysosome Fusion. *Frontiers in Immunology*.

Jeremy N. Timmis, M. A. (2004). Endosymbiotic gene transfer: organelle genomes forge eukaryotic chromosomes. *Nature Reviews Genetics*, 123-135.

Jessica M. Nelson, D. A.-W. (2020). Symbiotic cyanobacteria communities in hornworts across time, space, and host species. *bioRxiv* .

Jiang, Y., Wang, C., Ma, R., Zhao, Y., Ma, X., Wan, J., Li, C., Chen, F., Fang, F., & Li, M. (2021). Aquaporin 1 mediates early responses to osmotic stimuli in endothelial cells via the calmodulin pathway.

FEBS Open Bio, 11(1), 75–84. https://doi.org/10.1002/2211-5463.13020

Jing Yan Li, C. F. (2005). New symbiotic hypothesis on the origin of eukaryotic flagella. *Naturwissenschaften*, 305.

Johannes M. Keegstra, F. C. (2022). The ecological roles of bacterial chemotaxis. *Nature Reviews Microbiology*, 491–504.

Johnson, M. D. (2011). Acquired phototrophy in ciliates: a review of cellular interactions and structural adaptations. *The Journal of Ekaryotic Microbiology*, 185-195.

Jong Im Kim, G. T. (2022). Gene loss, pseudogenization, and independent genome reduction in non-photosynthetic species of Cryptomonas (Cryptophyceae) revealed by comparative nucleomorph genomics. *BMC Biology*.

Jürgen F. H. Strassert, I. I. (2021). A molecular timescale for eukaryote evolution with implications for the origin of red algal-derived plastids. *Nature Communications volume*.

Karnkowska, A., Yubuki, N., Maruyama, M., Yamaguchi, A., Kashiyama, Y., Suzaki, T., Keeling, P. J., Hampl, V., & Leander, B. S. (2023). Euglenozoan kleptoplasty illuminates the early evolution of photoendosymbiosis. Proceedings of the National Academy of Sciences, 120(12). https://doi.org/10.1073/pnas.2220100120

Khandia, R., Dadar, M., Munjal, A., Dhama, K., Karthik, K., Tiwari, R., Yatoo, Mohd. I., Iqbal, H. M. N., Singh, K. P., Joshi, S. K., & Chaicumpa, W. (2019). A Comprehensive Review of Autophagy and Its Various Roles in Infectious, Non-Infectious, and Lifestyle Diseases: Current Knowledge and Prospects for Disease Prevention, Novel Drug Design, and Therapy. Cells, 8(7), 674.

https://doi.org/10.3390/cells8070674KASTING, J. F. (1993). Earth's Early Atmosphere. *Science*, 920-926.

Kirsty M. Hooper, E. J. (2022). V-ATPase is a universal regulator of LC3-associated phagocytosis and non-canonical autophagy. *Journal of Cell Biology*.

Knoll, A. H. (2012, January 20). *Lynn Margulis, 1938–2011*. Retrieved from Proceedings of the National Acedemy of Sciences: https://www.pnas.org/doi/full/10.1073/pnas.1120472109

Laetz, E. M. J., & Wägele, H. (2017). Chloroplast digestion and the development of functional kleptoplasty in juvenile Elysia timida (Risso, 1818) as compared to short-term and non-chloroplast-retaining sacoglossan slugs. PLOS ONE, 12(10), e0182910. https://doi.org/10.1371/journal.pone.0182910

Levandowsky, M. (2012). Chapter 49 - Physiological Adaptations of Protists . In N. Sperelakis, *Cell Physiology Source Book (Fourth Edition)* (pp. 873-890). Academic Press.

Linda Bonen, W. F. (1975). On the Prokaryotic Nature of Red Algal Chloroplasts. *Proceedings of the National Academy of Science*, 2310-2314.

Ling Fang, F. L.-H.-J. (2017). Evolution of the Chlorophyta: Insights from chloroplast phylogenomic analyses. *Journal of Systematics and Evolution*, 322-332.

Liting Lim, G. I. (2010). The evolution, metabolism and functions of the apicoplast. *Philosophical Transactions of the Royal Society B*, 749–763.

Lucie Vančurová, J. M. (2021). Choosing the Right Life Partner: Ecological Drivers of Lichen Symbiosis. *Frontiers in Microbiology*.

Lunine, J. I. (1998). 11 - The Hadean Earth. In J. I. Lunine, *Earth - Evolution of a Habitable World* (pp. 115-133). Cambridge: Cambridge University Press.

M. Vesteg, R. V. (2009). On the origin of chloroplasts, import mechanisms of chloroplast-targeted proteins, and loss of photosynthetic ability — review. *Folia Microbiologica*, 303–321.

Malihe Mehdizadeh Allaf, H. P. (2022). Cyanobacteria: Model Microorganisms and Beyond. *Microorganisms*, 696.

Marie-France Carlier, S. S. (2017). Global treadmilling coordinates actin turnover and controls the size of actin networks. *nature reviews molecular cell biology*, 389–401.

Marina Santic, R. A. (2008). Acquisition of the Vacuolar ATPase Proton Pump and Phagosome Acidification Are Essential for Escape of Francisella tularensis into the Macrophage Cytosol. *Infection and Immunity*, 2671-2677.

Marko Kaksonen, C. P. (2006). Harnessing actin dynamics for clathrin-mediated endocytosis. *Nature Reviews Molecular Biology*, 404-414.

Matthias Corrotte, S. C.-G.-F. (2006). Dynamics and Function of Phospholipase D and Phosphatidic Acid During Phagocytosis. *Traffic*, 365-377.

McKee-Muir, O. C., & Russell, R. C. (2017). Mechanistic Target of Rapamycin. In Autophagy: Cancer, Other Pathologies, Inflammation, Immunity, Infection, and Aging (pp. 231–250). Elsevier. https://doi.org/10.1016/B978-0-12-812146-7.00009-3

McVey, M., Khodaverdian, V. Y., Meyer, D., Cerqueira, P. G., & Heyer, W.-D. (2016). Eukaryotic DNA Polymerases in Homologous

Recombination. Annual Review of Genetics, 50(1), 393–421. https://doi.org/10.1146/annurev-genet-120215-035243

Menolfi, D., & Zha, S. (2020). ATM, ATR and DNA-PKcs kinases—the lessons from the mouse models: inhibition ≠ deletion. Cell & Bioscience, 10(1), 8. https://doi.org/10.1186/s13578-020-0376-xMichael Borg, S. A.-H. (2023). Red macroalgae in the genomic era. *New Phytologist*, 471-488.

Michael P Collins, M. F. (2020). Regulation and function of V-ATPases in physiology and disease. *Biochimica et Biophysica Acta (BBA) - Biomembranes*.

Michelle M Leger, M. K. (2017). Organelles that illuminate the origins of Trichomonas hydrogenosomes and Giardia mitosomes. *Nature ecology & evolution*, 92.

Ming He, J. W. (2019). Genetic basis for the establishment of endosymbiosis in Paramecium. *The ISME Journal volume*, 1360–1369.

Minjie Hu, Y. B. (2023). Coral–algal endosymbiosis characterized using RNAi and single-cell RNA-seq. *Nature Microbiology*, 1240-1251.

Monique Turmel, M.-C. G. (2009). The Chloroplast Genomes of the Green Algae Pyramimonas, Monomastix, and Pycnococcus Shed New light on the Evolutionary History of Prasinophytes and the Origin of the Secondary Chloroplasts of Euglenids. *Molecular Biology and Evolution*, 631–648.

Morgan E. Milton, S. W. (2016). Replication and maintenance of the Plasmodium falciparum apicoplast genome. *Molecular and Biochemical Parasitology*, 56-64.

Natalya Yutin, M. Y. (2009). The origins of phagocytosis and eukaryogenesis. *Biology Direct*.

Nedeljka Rosic, E. Y.-K.-G. (2015). Unfolding the secrets of coral–algal symbiosis. *ISME J*, 844–856.

Netra Pal Meena, A. R. (2017). Chemotactic network responses to live bacteria show independence of phagocytosis from chemoreceptor sensing. *eLife*.

Nicolas Tapon, A. H. (1997). Rho, Rac and Cdc42 GTPases regulate the organization of the actin cytoskeleton. *Current Opinion in Cell Biology*, 86-92.

Nobuko Arisue, T. H. (2015). Phylogeny and evolution of apicoplasts and apicomplexan parasites. *Parasitology International*, 254-259.

Nur Akmal Solehah Din, '. S.-S.-S. (2022). Brown Algae as Functional Food Source of Fucoxanthin: A Review. *MDPI*.

Oborník, M. (2019). Endosymbiotic Evolution of Algae, Secondary Heterotrophy and Parasitism. *Biomolecules*, 266.

Olejarz, J., Iwasa, Y., Knoll, A. H., & Nowak, M. A. (n.d.). The Great Oxygenation Event as a consequence of ecological dynamics modulated by planetary change. https://doi.org/10.1038/s41467-021-23286-7

Olivier De Clerck, K. A. (2012). Chapter Two - Diversity and Evolution of Algae: Primary Endosymbiosis. In G. Piganeau, *Advances in Botanical Research* (pp. 55-86). Academic Press.

Olson, J. M. (2006). Photosynthesis in the Archean Era. *Photosynthesis Research*, 109-117.

Onuma, R. (2020, June 12). *Thievery as a lesson to become photosynthetic eukaryotes? –a transient endosymbiosis in a kleptoplastic dinoflagellate*. Retrieved from Behind the Paper - Microbiology: https://microbiologycommunity.nature.com/posts/thievery-as-a-lesson-to-become-photosynthetic-eukaryotes-a-transient-endosymbiosis-in-a-kleptoplastic-dinoflagellate

Ouyang, K. J., Woo, L. L., & Ellis, N. A. (2008). Homologous recombination and maintenance of genome integrity: Cancer and aging through the prism of human RecQ helicases. Mechanisms of Ageing and Development, 129(7–8), 425–440. https://doi.org/10.1016/j.mad.2008.03.003

Pannunzio, N. R., Watanabe, G., & Lieber, M. R. (2018). Nonhomologous DNA end-joining for repair of DNA double-strand breaks. Journal of Biological Chemistry, 293(27), 10512–10523. https://doi.org/10.1074/jbc.TM117.000374

Patrick J. Keeling, J. P. (2017). Endosymbiosis: the feeling is not mutual. *Journal of Theoretical Biology*, 75-79.

Patrick J. Keeling, V. M. (2021). Corallicolids: The elusive coral-infecting apicomplexans. *PLoS Pathogens*.

Paulina Kuczynska, M. J.-R. (2015). Photosynthetic Pigments in Diatoms. *MDPI*, 5847-5881.

Pekka Lappalainen, T. K.-L. (2022). Biochemical and mechanical regulation of actin dynamics. *Nature Reviews Molecular Cell Biology* , 836-852.

Pelletreau, K. N., Weber, A. P. M., Weber, K. L., & Rumpho, M. E. (2014). Lipid Accumulation during the Establishment of Kleptoplasty in Elysia chlorotica. PLoS ONE, 9(5), e97477. https://doi.org/10.1371/journal.pone.0097477

Patricia Sánchez-Baracaldo, G. B. (2022). Cyanobacteria and biogeochemical cycles through Earth history. *Trends in Microbiology*, 143-157.

Peter Civáň, P. G. (2014). Analyses of Charophyte Chloroplast Genomes Help Characterize the Ancestral Chloroplast Genome of Land Plants. *Genome Biology and Evolution*, 897-911.

Qing Tang, K. P. (2020). A one-billion-year-old multicellular chlorophyte. *Nature Ecology & Evolution*, 543-549.

Qingxian Lu, Q. L. (2010). Regulation of phagocytosis by TAM receptors and their ligands. *Frontiers in Biology*, 227–237.

Rafael I Ponce-Toledo, D. M.-G. (2018). Secondary Plastids of Euglenids and Chlorarachniophytes Function with a Mix of Genes of Red and Green Algal Ancestry . *Molecular Biology and Evolution*, 2198-2204.

Rafferty, J. P. (2023, 9 14). *Hadean Eon*. Retrieved from Encyclopedia Britannica: https://www.britannica.com/science/Hadean-Eon

Rafiqi, Ab. M., Polo, P. G., Milat, N. S., Durmuş, Z. Ö., Çolak-Al, B., Alarcón, M. E., Çağıl, F. Z., & Rajakumar, A. (2022). Developmental Integration of Endosymbionts in Insects. Frontiers in Ecology and Evolution, 10. https://doi.org/10.3389/fevo.2022.846586

Rebecca S. Meaney, S. H. (2020). Designer endosymbionts: Converting free-living bacteria into organelles. *Current Opinion in Systems Biology*, 41-50.

Richard G. Dorrell, C. J. (2015). Integration of plastids with their hosts: Lessons learned from dinoflagellates. *Proceedings of the National Academy of Science*, 10247-10254.

Romain Derelle, P. L.-G. (2016). A Phylogenomic Framework to Study the Diversity and Evolution of Stramenopiles (=Heterokonts). *Molecular Biology and Evolution*, 2890-2898.

Roni Levin-Konigsberg, A. R. (2021). A guide to measuring phagosomal dynamics. *The Febs Journal*, 1412-1433.

Rosa Isabel Figueroa, I. B. (2009). The Life History and Cell Cycle of Kryptoperidinium foliaceum, A Dinoflagellate with Two Eukaryotic Nuclei. *Protist*, 285-300.

Ross F Waller, L. K. (2017). Plastid complexity in dinoflagellates: a picture of gains, losses, replacements and revisions. In *Advances in Botanical Research vol. 84* (pp. 105-143). Elsevier.

Roth, M. S. (2014). The engine of the reef: photobiology of the coral–algal symbiosis. *Frontiers in Microbiology*.

Ryoma Kamikawa, G. T. (2015). Plastid Genome-Based Phylogeny Pinpointed the Origin of the Green-Colored Plastid in the Dinoflagellate Lepidodinium chlorophorum . *Genome Biology and Evolution*, 1133–1140.

Sagan, D. (2012). Lynn Margulis: The Life and Legacy of a Scientific Rebel. Chelsea Green Publishing.

Sagan, L. (1967). On the origin of mitosing cells. Journal of Theoretical Biology, 225-274.

Sang-Moo Lee, C.-M. R. (2021). Algae as New Kids in the Beneficial Plant Microbiome. Frontiers in Plant Science.

Sarah E. Glass, R. M. (2023). Chloroplast genome evolution and phylogeny of the early-diverging charophycean green algae with a focus on the Klebsormidiophyceae and Streptofilum. Journal of Phycology.

Sarah Méresse, M. F. (2020). Fucoxanthin, A Marine-Derived Carotenoid from Brown Seaweeds and Microalgae: A Promising Bioactive Compound for Cancer Therapy. International Journal of Molecular Science.

Saunders GW, H. M. (2004). The origin of red algae: Implications for plastid evolution. Proceedings of the National Academy of Sciences, 5438-5443.

Sato, N., Yoshitomi, T., & Mori-Moriyama, N. (2020). Characterization and Biosynthesis of Lipids in Paulinella micropora MYN1: Evidence for Efficient Integration of Chromatophores into Cellular Lipid Metabolism. Plant and Cell Physiology, 61(5), 869–881. https://doi.org/10.1093/pcp/pcaa011

Shenglan Li, T. N. (2005). Phylogenomic Analysis Identifies Red Algal Genes of Endosymbiotic Origin in the Chromalveolates. Molecular Biology and Evolution, 663-674.

Shibbir Ahmed Khan, M. A. (2023). The Phospholipase A2 Superfamily: Structure, Isozymes, Catalysis, Physiologic and Pathologic Roles. International Journal of Molecular Sciences.

Shigekatsu Suzuki, R. M. (2022). What Happened before Losses of Photosynthesis in Cryptophyte Algae? Molecular Bilology Evolution.

Shigekatsu Suzuki, Y. H.-I. (2016). Plastid genome sequences of Gymnochlora stellata, Lotharella vacuolata, and Partenskyella glossopodia reveal remarkable structural conservation among chlorarachniophyte species. Journal of Plant Research, 581-590.

Shinichiro Maruyama, T. S. (2011). Eukaryote-to-eukaryote gene transfer gives rise to genome mosaicism in euglenids. BMC Ecology and Evolution.

Shih, P. M. (2015). Cyanobacterial Evolution: Fresh Insight into Ancient Questions. Current Biology, 25(5), R192–R193. https://doi.org/10.1016/j.cub.2014.12.046

Shunsuke Hirookaa, T. I. (2022). Life cycle and functional genomics of the unicellular red alga Galdieria for elucidating algal and plant evolution and industrial use. Proceedings of the National Academy of Science.

Sibbald, S. J., & Archibald, J. M. (2020). Genomic Insights into Plastid Evolution. Genome Biology and Evolution, 12(7), 978–990. https://doi.org/10.1093/gbe/evaa096

Sloan, D. B., Warren, J. M., Williams, A. M., Wu, Z., Abdel-Ghany, S. E., Chicco, A. J., & Havird, J. C. (2018a). Cytonuclear integration and co-evolution. Nature Reviews Genetics, 19(10), 635–648. https://doi.org/10.1038/s41576-018-0035-9

Sofie E. Voerman, A. R. (2022). Red algae acclimate to low light by modifying phycobilisome composition to maintain efficient light harvesting. BMC Biology.

SooNee Tan, D. G. (2021). Properties of Plasmodium falciparum with a Deleted Apicoplast DNA Gyrase. Antimicrob Agents Chemotherapy.

Sproles, A. E., Kirk, N. L., Kitchen, S. A., Oakley, C. A., Grossman, A. R., Weis, V. M., & Davy, S. K. (2018). Phylogenetic characterization of transporter proteins in the cnidarian-dinoflagellate symbiosis. Molecular Phylogenetics and Evolution, 120, 307–320. https://doi.org/10.1016/j.ympev.2017.12.007

Steven L. Porter, G. H. (2011). Signal processing in complex chemotaxis pathways. Nature Reviews Microbiology, 153-165.

Szczepanowska, J. (2009). Involvement of Rac/Cdc42/PAK pathway in cytoskeletal rearrangements. Acta Biochimica Polonica.T. Martin Embley, W. M. (2006). Eukaryotic evolution, changes and challenges. *Nature*, 623-630.

Takahashi, T. (2016). Simultaneous Evaluation of Life Cycle Dynamics between a Host Paramecium and the Endosymbionts of Paramecium bursaria Using Capillary Flow Cytometry. Scientific Reports.

Tamar L. Goulet, D. G. (2021). Climate Change Leads to a Reduction in Symbiotic Derived Cnidarian Biodiversity on Coral Reefs. Frontiers in Ecology and Evolution.

Tingting Xiang, E. L. (2020). Symbiont population control by host-symbiont metabolic interaction in Symbiodiniaceae-cnidarian associations. Nature Communications.

Toshiyuki Takahashi, T. N. (2016). Delineation of six species of the primitive algal genus Glaucocystis based on in situ ultrastructural characteristics. Scientific Reports.

Tove M. Gabrielsen, M. A.-K.-T. (2011). Genome Evolution of a Tertiary Dinoflagellate Plastid. PLoS One.

Timmis, J. N., Ayliffe, M. A., Huang, C. Y., & Martin, W. (2004). Endosymbiotic gene transfer: organelle genomes forge eukaryotic chromosomes. Nature Reviews Genetics, 5(2), 123–135. https://doi.org/10.1038/nrg1271

Tooze, S. A., Abada, A., & Elazar, Z. (2014). Endocytosis and Autophagy: Exploitation or Cooperation? Cold Spring Harbor Perspectives in Biology, 6(5), a018358–a018358. https://doi.org/10.1101/cshperspect.a018358

Tsuji, Y. (2023). Kleptoplasty Relies on a Host-Derived Component in the Euglenid Protist, Rapaza viridis. *Plant And Cell Physiology*, 64(9), 1079–1081. https://doi.org/10.1093/pcp/pcad070

Trustees of Dartmouth College. (n.d.). Lynn Margulis. The Montgomery Fellows Program.U. Kutschera, K. N. (2005). Endosymbiosis, cell evolution, and speciation. *Theory in Biosciences*, 1-24.

Ugo Cenci, S. J. (2018). Nuclear genome sequence of the plastidlacking cryptomonad Goniomonas avonlea provides insights into the evolution of secondary plastids. *BMC Biology*.

Ulrike Neumann, F. D.-S. (2019). Fucoxanthin, A Carotenoid Derived from Phaeodactylum tricornutum Exerts Antiproliferative and Antioxidant Activities In Vitro. *Antioxidants*, 183.

University of Houston. (n.d.). *Faculty Profile*. Retrieved from University of Houston: https://uh.edu/nsm/biology-biochemistry/people/profiles/george-fox/

University of Ottawa . (n.d.). *University of Ottawa* . Retrieved from University of Ottawa : https://www.uottawa.ca/faculty-science/professors/linda-bonen

Valentin Jaumouillé, C. M. (2020). Physical Constraints and Forces Involved in Phagocytosis. *Frontiers in Immunology*.

Valeria Villanova, C. S. (2021). Mixotrophy in diatoms: Molecular mechanism and industrial potential. *Physiologia Plantarum*, 603-611.

Van Steenkiste, N. W. L., Stephenson, I., Herranz, M., Husnik, F., Keeling, P. J., & Leander, B. S. (2019). A new case of kleptoplasty in animals: Marine flatworms steal functional plastids from diatoms. *Science Advances*, 5(7). https://doi.org/10.1126/sciadv.aaw4337

Varsha Mathur, M. K. (2019). Multiple Independent Origins of Apicomplexan-Like Parasites. *Current Biology*.

Varsha Mathur, W. K. (2021). Phylogenomics Identifies a New Major Subgroup of Apicomplexans, Marosporida class nov., with Extreme Apicoplast Genome Reduction. *Genome Biology and Evolution*.

Vincent Daubin, G. J. (2016). Horizontal Gene Transfer and the History of Life. *Cold Spring Harb Perspect Biology*.

W. Benz, W. S. (1986). The origin of the moon and the single-impact hypothesis I. *Icarus*, 515-535.

W.D.P. Stewart, P. R. (1983). Cyanobacteria-eukaryotic plant symbioses. *Annales de l'Institut Pasteur / Microbiologie*, 205-228.

Waldan K Kwong, J. D. (2019). A widespread coral-infecting apicomplexan with chlorophyll biosynthesis genes. *Nature*, 103-107.

Wernegreen, J. J. (2012). Endosymbiosis. *Current Biology*, 555-561.

William F. Martin, S. G. (2015). Endosymbiotic theories for eukaryote origin. *Phylosophical Transactions of the Royal Society B*.

William Martin, J. B. (2008). Hydrothermal vents and the origin of life. *Nature Reviews Microbiology*, 805-814.

William Martin, K. K. (2010). Annotated English translation of Mereschkowsky's 1905 paper 'Über Natur und Ursprung der Chromatophoren imPflanzenreiche'. *European Journal of Phycology*, 287-295.

William Martin, M. M. (1998). The hydrogen hypothesis for the first eukaryote. *Nature*, 37-41.

Xi Li, Z. H. (2021). Large Phylogenomic Data sets Reveal Deep Relationships and Trait Evolution in Chlorophyte Green Algae. *Genome Biology and Evolution.*

Xiang, T., Lehnert, E., Jinkerson, R. E., Clowez, S., Kim, R. G., DeNofrio, J. C., Pringle, J. R., & Grossman, A. R. (2020). Symbiont population control by host-symbiont metabolic interaction in Symbiodiniaceae-cnidarian associations. Nature Communications, 11(1), 108. https://doi.org/10.1038/s41467-019-13963-z

Yaacov Davidov, E. J. (2009). Predation between prokaryotes and the origin of eukaryotes. *BioEssays.*

Yamamoto, H., Zhang, S., & Mizushima, N. (2023). Autophagy genes in biology and disease. Nature Reviews Genetics, 24(6), 382–400. https://doi.org/10.1038/s41576-022-00562-w

Yan Yang, 1. G.-D. (2022). Receptor for advanced glycation end-products (RAGE) mediates phagocytosis in nonprofessional phagocytes. Communications Biology.

Yuqian Tang, D. Q.-P. (2023). Diurnal switches in diazotrophic lifestyle increase nitrogen contribution to cereals. Nature Communications.

Yutaka Kato, N. I. (2009). Metabolic Control Between the Symbiotic Chlorella and the Host Paramecium. In S. Link, Endosymbionts in Paramecium (pp. 57–82). Springer Link.

Yuuki Kodama, M. F. (2022). Endosymbiotic Chlorella variabilis reduces mitochondrial number in the ciliate Paramecium bursaria. Scientific Reports.

Zhao, B., Rothenberg, E., Ramsden, D. A., & Lieber, M. R. (2020). The molecular basis and disease relevance of non-homologous DNA end joining. *Nature Reviews Molecular Cell Biology*, 21(12), 765–781. https://doi.org/10.1038/s41580-020-00297-8